Being-in-Creation

gROUNDWORKS|

ECOLOGICAL ISSUES IN PHILOSOPHY AND THEOLOGY

Forrest Clingerman and Brian Treanor, *series editors*

Being-in-Creation

Human Responsibility
in an Endangered World

EDITED BY

Brian Treanor,
Bruce Ellis Benson,
and Norman Wirzba

Fordham University Press | *New York 2015*

Fordham University Press has no responsibility for the persistence or accuracy of URLs for external or third-party Internet websites referred to in this publication and does not guarantee that any content on such websites is, or will remain, accurate or appropriate.

Fordham University Press also publishes its books in a variety of electronic formats. Some content that appears in print may not be available in electronic books.

Visit us online at www.fordhampress.com.

Library of Congress Cataloging-in-Publication Data

Being-in-creation : human responsibility in an endangered world / edited by Brian Treanor, Bruce Ellis Benson, and Norman Wirzba.
    pages cm. — (Groundworks : ecological issues in philosophy and theology)
    Includes bibliographical references and index.
    ISBN 978-0-8232-6499-5 (hardback) — ISBN 978-0-8232-6500-8 (paper)
    1. Human ecology—Religious aspects—Christianity. 2. Creation. 3. Theological anthropology—Christianity. I. Treanor, Brian, editor. II. Benson, Ellis, (date)–editor. III. Wirzba, Norman, editor.
    BT695.5.B43 2015
    261.8′8—dc23

2014040683

Printed in the United States of America

17 16 15   5 4 3 2 1

First edition

Contents

Acknowledgments

A collection like this one is, inevitably, in debt to the work of more than just the contributors to and editors of the final volume. The editors would like to express their gratitude to the following individuals, institutions, and organizations.

Helen Tartar and Thomas Lay, of Fordham University Press, carefully oversaw the process of bringing this collection to press. Just before the final version of the manuscript was to be submitted to Fordham, we received the shocking news that Helen had passed away after a tragic car accident. It would be difficult to overstate the influence Helen had on publishing continental philosophy in the United States, including works with either, or both, religious and environmental overtones. We will miss her sensitivity to the directions in which scholarship was moving, the care with which she nurtured new authors, the collegial friendship she shared with established scholars, her gentle spirit, and her keen wit. Helen's legacy, which is substantial, lives on in the many books she shepherded to print and, we hope, in this one.

The Bellarmine College of Liberal Arts at Loyola Marymount University generously supported the 2012 meeting of the Society for Continental Philosophy and Theology (SCPT), which saw the first versions of many of the papers presented here. Donald Boyce was a great help in the preparation of the final manuscript for publication.

His conscientious proofreading and formatting helped to streamline the work of editing during the final phases of the project.

A few of the essays in this volume have previously appeared in print or in public presentation, and the editors would like to express their gratitude to the following entities for their willingness to allow the reproduction of those pieces in this volume: "Creation and the Glory of Creatures," by Janet M. Soskice, appeared in *Modern Theology* 29, no. 2 (April 2013): 172–85; "Creation, Creativity and Creatureliness: The Wisdom of Finite Existence" was a lecture delivered by Dr. Rowan Williams at St. Theosevia Centre for Christian Spirituality at Oxford on April 23, 2005; Norman Wirzba's "The Art of Creaturely Life" appeared in *Pro Ecclesia* 22, no.1 (Winter 2013).

Being-in-Creation

Introduction

The Human Place
in the Natural World

David Brian Treanor

> I come into the presence of still water.
> And feel above me the day-blind stars
> waiting with their light. For a time
> I rest in the grace of the world, and am free.
> —Wendell Berry, "The Peace
> of Wild Things" (1968)

Where Do We Belong?

Philosophers, theologians, poets, and storytellers have all wondered about their place in the universe, about *our* place—the place of humans—in the wider cosmos or in creation. The very fact that we feel compelled to ask the questions about where we fit and how we belong implies something significant about the human condition: that we don't fit—or *feel* we don't fit—into the wider fabric of the world. We find ourselves suspended, like Ezekiel, between heaven and Earth, challenged to find our way in a world that is both familiar and foreign, one that both fits and chafes.

Indeed, to ask after the "human place in the natural world" is to open certain controversial lines of inquiry. Are humans—and with them human language, culture, technology, and civilization—part of the natural order or distinct from it? If the latter is true, must we posit a human place elsewhere than in the natural world, a sphere

1

in which humans might be placed, perhaps more appropriately? Is the human place in the natural world something we are describing, a fact about our own nature and our place in the wider world? Or, alternatively, is the human place in the natural world something for which we search, something we need to find, or make, since we aren't at home here easily or naturally?

On some level, of course, we do fit in nature, and thinkers like Gary Snyder forcefully insist on human belonging in the natural world: "Nature is not a place to visit, it is *home*."[1] But the fact that we find ourselves here, thrown into this world and into this nature without any clear sense of what it all *means*, leaves us with some lingering anxiety that there must be something more, that this cannot be all there is. Wallace Stegner tells us that "home" is, ultimately, the concern of those who either are, or else feel themselves to be, homeless: "Home is a notion that only the nations of homeless fully appreciate and only the uprooted comprehend."[2] However, most people seem to be torn by two different impulses related to their ultimate belonging in nature: on the one hand, they are gripped by some sense of the exceptional and, perhaps, superior aspects of human nature that would distinguish us from the rest of nature; on the other hand, they cannot deny the thoroughgoing ordinariness of our constitution and our fundamental kinship with all other living things.

The seeming specialness and importance of human beings are widely endorsed in both thought and act. Juxtaposed with other beings in the world, we call attention to our particular status as rational animals (Aristotle's *zoon logon echon*), or to the claim that we are beings composed of an immaterial soul in addition to a material body (as Descartes argues), or the assumption that we are intelligent life on a planet with exceedingly rare conditions conducive to the evolution of complex life forms (as held by versions of the Rare Earth hypothesis), or beings with singular capacities for creativity, humor, art, and beauty. However, one of the oldest—and arguably one of the most influential—of the various ways of distinguishing humans from the rest of the cosmos (and one for which this collection has a certain affinity) is the claim that humans, unique among all other beings in creation, were made in the *imago dei*.

Of course, the notion of the *imago dei* has generated a long history of philosophical, theological, and environmental controversy. In attacking Gnosticism, St. Irenaeus distinguishes between the image (*selem*) and likeness (*demut*) of God, arguing that, in a postlapsarian world, we retain the former but have lost the latter.[3] Augustine struggles with the implication that the *imago dei* would impute some changeability to God's unchanging nature, ultimately coming to understand that the term did not suggest similarity in form or substance.[4] Aquinas locates the *imago dei* in our rationality and claims it may be found to a greater or lesser extent in different individuals (corresponding to their knowledge and love of God) or types (thus, angels have it to a greater degree than humans).[5] Karl Barth posits a fundamental relational aspect to the *imago dei*, arguing that it can be found in the relationship between men and women and in the relationship between individuals and God.[6]

In terms of modern environmental thinking, Lynn White's celebrated, maligned, and endlessly referenced essay "The Historical Roots of our Ecologic Crisis"[7] places the blame for our present ecological crises squarely on the Genesis account of creation and its subsequent interpretations, taking issue with the apparent claims of superiority implicit in doctrines of *imago dei* and with exhortations to domination and exploitation based on it: "Let us make man in our image, after our likeness. . . . And God blessed them, and said to them, 'Be fruitful and multiply, and fill the earth and subdue it; and have dominion (*kibbes*) over the fish of the sea and over the birds of the air and over every living thing that moves upon the earth'" (Gen. 1:26–28). White argues that Christianity, "the most anthropocentric religion the world has ever seen,"[8] has led us squarely into the environmental crisis by suggesting that we have absolute dominion over the Earth, including the right to use it in a gluttonous or profligate way. And this charge was leveled in 1967, before the changes in landscapes and ecosystems associated with over seven billion humans (from urbanization, agriculture, transportation, and so on), before genetic engineering, and well before we came to realize the full significance of anthropogenic climate change.

In whatever way we justify them, claims of inherent, natural, or ontological exceptionalism on behalf of human beings tend to set us apart from the rest of the world, leading to distinctions between culture and nature, civilization and wilderness, life and inert matter, human and animal, and, all too often, the intrinsically valuable and merely useful. Such facile dualisms are worthy of critical evaluation on their own, but there are significant practical problems that issue from these conceptual differentiations. Under the influences of such divisive juxtapositions it is not at all difficult to see how we come to disparage—or at the very least distrust—materiality itself,[9] distinguishing the body from the soul and, perhaps, even coming to view the former as the prison house of the latter.[10] Likewise, our lauding of reason, the built environment, and human distinctiveness leads us to denigrate emotion, the earth, and our basic animality.

However, while we tend to assert human superiority in both our philosophies and in our practices, these assertions of intrinsic superiority are in deep tension with an increasing awareness and appreciation of our thoroughgoing ordinariness, and perhaps even our insignificance. Our ordinariness is evident in our biology: we are animals, evolved by the same processes as other animals, living on the same planet under largely the same conditions. We live with other plant and animal beings—to say nothing of bacteria, protozoa, fungi, and other forms of life—alongside whom we flourish or fail to flourish in an interconnected web of life. Increasingly, we run the risk of shared dooms or apocalypses, such as runaway climate change. True, we possess certain rare capacities (e.g., language); but we also lack others (e.g., echolocation, the ability to spin webs). Moreover, we now understand that many of the traits we once thought characteristic of only human beings are actually shared, in varying but often significant degrees, by other animals. Even the traditions that advanced the idea of human beings being made (uniquely) *imago dei* turn out to be polyvocal with respect to creaturely value. The Book of Job acknowledges human value but also the value of all other creatures—naming lions, mountain goats, ravens, wild asses, wild oxen, ostriches, horses, and hawks, as well as the terrifying otherness of the "Behemoth" and the "Leviathan." The perspective here is neither anthropocentric or biocentric, but theocentric; and it places human life within a wide, deeply complex, and terrifying yet God-loved world.[11]

Our insignificance is even starker and more unsettling: consider the span of Methuselah against the backdrop of the roughly fourteen billion years since the Big Bang, or the reach of an "empire on which the sun never sets" against images of Hubble's various "deep field" photographs, which reveal an observable universe with more than 125 billion galaxies and some 3×10^{21} suns. Our greatest works are as nothing when seen against the backdrop of the vast expanses of interstellar space and time. Michelangelo's frescoes in the Sistine Chapel are unlikely to survive as long as the paintings of Lascaux, which, despite a seventeen-thousand-year history, are themselves turning out to be strikingly fragile, perishable, and finite. Translatable, transcribable, and transmissible as they are, the collected dialogues of Plato, the Bhagavad-Gita, the Bible, Bach's cello suites, and similar works may enjoy a somewhat longer lifespan; but they too are ultimately perishable. On a long enough time horizon, the greatest human works—the pyramids of Egypt, the Great Wall of China, the skylines of our most mighty contemporary cities, and the artistic, cultural, and intellectual treasures of every civilization—stand as "two vast and trunkless legs of stone" against the "lone and level sands" of the cosmos, and that only briefly before they too fall to the Second Law of Thermodynamics and are reduced to nothing.[12] Stardust we are, and to stardust we shall return. Amen. The universe yawns and takes another spin.[13]

However, both of these perspectives—human uniqueness and superiority and human ordinariness and perhaps insignificance—lead to problematic conclusions. Strident claims of superiority, whether due to some apparently exceptional trait, some unique or essential characteristic, or some supernatural watermark, generally come at the cost of estrangement from, and all too often conflict with, the natural world. In contrast, when we insist absolutely on the thoroughly unexceptional character of our makeup and root ourselves firmly in the natural world alongside other beings, we perhaps do so at the cost of deriving any ultimate significance to either our culture or our nature.

Continental Philosophy: *Das Heimelige* and *das Unheimliche*

What might continental philosophy have to say about this state of affairs? Might we find another way of thinking about the human place in the world that avoids both meaning generated through the unreflective anthropocentrism and hubris associated with facile accounts of human superiority, as well as the tendency to despair and nihilism associated with a cold, dark, and ultimately meaningless universe in which there is nothing but matter and motion?[14] Is there an alternative perspective that embraces both our exceptionalism and our ordinariness, the transcendent and the earthy, the spiritual and the material, the cultural and the natural, the human and the nonhuman, the secular and the sacred, our unique gifts and our kinship with everything else, the wild alongside the pastoral and the urban? This is no small task, and there is certainly more than one way to go about it. However, several of the contributors to this volume suggest that such an alternative perspective can be found by embracing our *creatureliness*, interpreting our nature and our place in nature in a manner that recognizes both our difference from and our kinship with the rest of the natural world. Charles Darwin himself struggled with this tension. On the one hand, he exalts the "noble qualities" and "god-like intellect" of man—which, echoing Hamlet ("What a piece of work is man! How noble in reason, how infinite in faculty!"), caused him to suggest we had risen to the "very summit of the organic scale" and that we might rise still higher ("In action how like an angel! In apprehension how like a god!"). On the other hand, Darwin recognizes the "lowly" origins and kinship we share with all other living things on this planet:

We must, however, acknowledge, as it seems to me, that man with all his noble qualities, with sympathy which feels for the most debased, with benevolence which extends not only to other men but to the humblest living creature, with his god-like intellect which has penetrated into the movements and constitution of the solar system—with all these exalted powers—Man still bears in his bodily frame the indelible stamp of his lowly origin.[15]

It's often been suggested that one of the strengths of continental philosophy is its concern with wisdom and with questions of meaning. But certainly one of the most essential and unavoidable questions of meaning is the question of where we belong: a mark of wisdom is to understand one's nature and one's place in the larger scheme of things, as well as to act in a manner proper to that nature and place.

Though philosophers might debate the originary text, figure, or moment that gave rise to what is now called continental philosophy—whether Kant's critical philosophy, German Romanticism and Idealism, Kierkegaard and Nietzsche, or Husserl—we can still identify predominant themes in the traditions that have been handed down to us.

At first blush, continental philosophy appears to favor some version of the perspective that emphasizes human uniqueness, drawing attention to both qualities that distinguish us from the rest of nature and to the alienation or dislocation that makes us strangers to the environment in which we find ourselves. Thus, Heidegger draws our attention to both our state of thrownness (*Geworfenheit*) and the resulting *Angst* (anxiety, dread) we feel in the face of *unheimlich* (uncanny) displacements and disconnection.[16] *Geworfenheit* suggests not only our thrownness into a particular language, culture, tradition, and set of narratives, but surely also our thrownness into the natural world with its multitudinous dangers, demands, and distractions. It suggests that we are in some primordial or essential sense *dis*-placed, and that we never quite throw off the feeling of being not-at-home or not-yet-home—we remain "strangers in the land of Egypt" (Exod. 22:21). Thus, while some thinkers focus on Heidegger's analysis of being-toward-death (*Sein-zum-Tode*), thrownness and displacement are also clearly present at the root of the *Angst* and *Unheimlichkeit* that so interest twentieth-century continental philosophers.

Unlike fear, *Angst* is not the result of some definite threat, but rather the result of something indefinite, something like the feeling that one is not "at home," that one doesn't "fit" in the world or in the order of things, that one is displaced in a world that is, somehow, strange and foreign. Heidegger pursues similar themes in later works such as "Building, Dwelling, Thinking," in which he suggests dwelling is essential to being human, but also points out that building and dwelling are distinct.[17] Indeed, it should be obvious to those reflecting on questions of the human place in the natural world—or, indeed, in the built world—that homelessness cannot be reduced to the absence of a dwelling.[18] Heidegger was certainly not the first thinker to put his finger on the importance of dwelling and place. Nor was he the first to recognize the experience of displacement that seems to be a fundamental possibility, or perhaps even a universal experience, of the human condition (consider Kafka, Dostoevsky, Nietzsche, Kierkegaard, and many others). Heidegger did, however, exert an enormous influence over the course of continental philosophy. Thinking in his wake, various strands of existentialism and postmodernism take up similar concerns by addressing disturbance, disruption, disconnection, displacement, *différance*, and similar subjects.

However, while the uncanniness and displacement of *Geworfenheit* are undoubtedly a constant possibility for *Dasein*, might it be the case that philosophy has overplayed both the primordiality and the significance of these unsettled and unsettling moods and feelings? As poet Gary Snyder reminds us, if the world can seem at times to be a hostile and alien(ating) place, we do well to remember that such a perspective is only an aspect of a much grander vista:

I have a friend who feels sometimes that the world is hostile to human life—he says it chills and kills us. But how would we *be* were it not for this planet that provided our very shape? Two conditions—gravity and a livable temperature between freezing and boiling—have given us fluids and flesh. The trees we climb and the ground we walk on have given us five fingers and toes. The "place" (from the root *plat*, broad, spreading, flat) gave us far-seeing eyes, the streams and breezes gave us versatile tongues and whorly ears. The land gave us a stride, and the lake a dive. The amazement gave us our kind of mind. We should be thankful for that, and take nature's stricter lessons with some grace.[19]

Might we consider, further, the possibility not only of stoic resignation to the harsher realities of our world and our nature, or even a Nietzschean endorsement of them in "eternal return" (*ewige Wiederkunft*), but perhaps also a genuine gratitude for our lives and for the world in which we find ourselves, a world to which we are deeply connected, a world with which we have evolved and to which we are adapted, a world in which we can and do find ourselves profoundly at home?

Indeed, continental philosophy and theology have been far from univocal in endorsing a displaced view of humans in the natural world. Perhaps the seminal work on the human place in the world is Edward Casey's *Getting Back into Place*, which paints a much more complex and rich picture of the various ways in which humans live in and interact with natural spaces and places.[20] Without denying the possibility of certain sorts of homelessness and placelessness, Casey makes a powerful case for the fundamental primordiality of place, which is "prior to all things": "To exist at all . . . is to have a place."[21] Theological thinkers like Forrest Clingerman have argued that Christianity includes, among other things, a *hermeneutic* way of navigating between place and placelessness at the intersection of nature, space, and the sacred.[22] Implacement—the sort of environment in which we are, in a sense, at home—is a form of interaction

between humans and nature that mediates a conceptual manifestation of place or "nature," concrete and particular places, and the
existential manifestation of these two elements.[23] Place and home
are *interpretations* we give to certain manifestations of space rather
than an undeniable and basic fact of nature, lending credence to Ni-
gerian novelist Ben Okri's exhortation "To anyone who is homeless,
I say, find a home."[24]

Perhaps human creatureliness and implacement in the natural
world are quite, well, natural, and the anxiety of *Unheimlichkeit*
is only a possibility for a being that is already at home (*Zuhause*)
in a world it finds rather homey (*heimelige*)—that the uncanny experience of displacement is an interruption of a more primordial
implacement. If so, then the displacement is a shadow cast by a
more fundamental implacement, something preceding and making
possible a subsequent homecoming (*Heimkommen*).[25] Thus, the
questions are not "what is the human place in the natural world?"
or "where do we fit?" Instead, the key questions are "what does it
mean to be a creature at home in this world?" and "how ought we
think and act and live as the kind of creatures we actually are, in
this world we've been given and in which we make our home?" The
first sorts of questions assume that the human place in the natural
world is something like the place of a joist in a roof or the location of
the oatmeal in my pantry, a discrete position or niche—physical or
psychological—that, once found, can be crossed off our existential
"to do" list. The second sorts of questions, however, assume that
the human place in the natural world is anything but static. We have
a place here, to be sure, and we belong in that place naturally. But
being at home in the world requires embracing a certain way of being, a way of being that recognizes our creatureliness: its powers and
its limitations; its freedoms as well as its dependencies (including
the dependence on an almost incomprehensibly complex web of relationships); its capacities and its vulnerability; and its transcendent
yearning alongside its inextricably earthy and animal nature. In order to be at home in nature, we have to recognize our *own* nature.
And we have to act in light of that understanding.

The glory of being human is the ability to recognize a pattern of rightness and to honor it as a moral law. The horror of being human is the ability to violate that rightness, living out of season—doing violence to the other, perverting the most sacred human relationships, devastating the world in greed, overriding its rhythm, not in the name of necessity and charity, but in the compulsion of coveting.[26]

We are not gods, nor are we sovereign lords over nonhuman nature. We are one kind of creature among other kinds of creatures, at home—once we realize this and settle down to make ourselves at home—in a world of remarkable beauty, fecundity, and resilience. Seen in this light, the *Unheimlichkeit* aspects of our situation are perhaps the result of our own alienation from our creatureliness and from our place, as creatures, in the world. This alienation—and the consequent exploitative view of the world as a container of resources on which we may draw at will—may be among the first things awaiting the theological hope for the transformation of creation into a new heaven and earth. Perhaps then, if we are unable to shake the anxiety that comes with thrownness and responsibility, the lingering worry that things don't quite fit, there is yet hope that we can help to accomplish this transformation, that we can, following poets like Elliot and philosophers like Ricoeur, after all our explorations, "arrive where we started [our home], and know it for the first time [as home]."

The Essays

The essays in this collection represent a wide variety of approaches bringing continental philosophy and philosophy of religion to bear on the question of the human place in the (more-than-human) natural world. At the intersection of continental philosophy and theology, these essays view this question in terms of creatureliness and creation. The collection arises out of discussions that began at the 2012 meeting of the Society for Continental Philosophy and Theology, and these discussions are therefore shaped by the scholars who contributed to that gathering and to those who continued dialogue in its wake. This genesis and development give the essays a unity shaped by both topic and approach. One might note, therefore, that the essays collected here happen to address creation and creatureliness from a broadly Christian perspective. Other philosophical and theological approaches are, of course, possible. Among the traditions of continental philosophy, for example, there is much work done under the influence of Jewish thinkers, not only Levinas and Derrida, but Rosenzweig, Buber, and others as well. Thus, the essays in this collection represent only one aspect of a larger dialogue, one that could, and should, be expanded to include not only other traditions of the book (Judaism, Islam) but also accounts of human creatureliness found in non-Western religions and spirituality.

It is true that, following White's critique, we could point to religious figures (of diverse faiths) who stress human exceptionalism to the point of caricature and farce, almost implying that it is precisely the exercise of dominion—not reason, not language, not love—that constitutes the distinctive essence of humanity. However, as many people have pointed out in response to White, these are not the only religious voices addressing the environment. Whether or not Christianity imported anthropocentrism to modern science in the Enlightenment (as White contends) and whether or not mainstream manifestations of many religions have behaved in anthropocentric ways, there are numerous voices in diverse faiths seeking to develop, or to revive, a different account of creatureliness and a different account of the role of the human creature in creation. The authors in this volume are among these voices.

There are pressing reasons to rethink (or reconnect with) what it means to be human in the wider context of creation, for while the unprecedented environmental crises facing us may be partially mitigated by technological innovation or economic incentives, we will never live sustainably over the long haul unless and until we come to accept that we are just one type of creature among many fellow creatures, rather than omnipotent gods exercising capricious (and intemperate) dominion over the rest of the cosmos.

"Creation, Creativity, and Creatureliness" was a lecture delivered by Dr. Rowan Williams, the former archbishop of Canterbury. It introduces us to the thought of Father Sergei Bulgakov, in whose work on the wisdom of God Williams sees "some remarkably challenging and resourceful ideas for coping with that bizarre environment in which we live called modernity." Bulgakov's account of "holy wisdom" offers us a rare model for thinking together creation, creativity, and creatureliness, one through which Williams sees the possibility of restoring and reenergizing the image of God in humanity. Bulgakov challenges us to understand how God and the world are related in a continuity that is also an unfathomable difference and, furthermore, to understand the meaning of our own *imago dei* and the unique calling of human beings in a universe like ours, made by a God like this. God, argues Bulgakov, empties himself in *kenosis* as an outpouring of life into otherness, a relation of self-giving. We cannot understand divine wisdom without understanding divine self-giving: wisdom is kenotic.

Thus the human calling to creativity mimics this divine kenosis by adopting a perspective that "stands back" and "lets be." Such creativity is not merely passive, and it is not content with producing beauty: creativity makes it possible for the world to change, and does so with an eye to truth and justice. Seen in this light, work in economics, justice, and ethics may all partake in this creative, kenotic, and sophianic project. This ethical and creative project is, ultimately, what is going on in the church, even when it fails to keep up with this vision. But we must keep in mind the difference between "uncreated" wisdom and creativity and "created" wisdom and creativity. Our creatureliness entails an acceptance that we are not God, one that, paradoxically, allows us to take part in the divine creative project. Williams sees a connection between our denial of creatureliness, our denial of finitude, and the environmental crises that beset us. In the belief that technology will always find a way for us to have more, do more, consume more, Williams hears the faint echo of Adam and Eve's desire to be "as gods," and perhaps of Milton's Satan, who declares it "better to rule in Hell, than to serve in Heaven."

"Rowan Williams and Ecological Rationality" takes issue with Slavoj Žižek's announcement of the "death of ecology," which addresses the claim that there is a difference between culture and nature, comforting us with the idea that we can "fix" things by changing our actions so that they no longer harm nature. According to Žižek, human culture—including technology, pollution, and trash—is just as natural as anything else. Culture is nature. However, while Longbons agrees with Žižek that humans are, in a sense, natural, he takes issue with the complete effacing of the nature-culture difference and argues that the key perspective that allows us to affirm both the continuity and the difference between nature and culture is to view the world as creation and humans as creaturely. Building on the work of Rowan Williams, Longbons argues that God's redemptive plan extends to nonhuman nature, the land, and indeed the entire cosmos. Humans play a part in this redemptive plan, but not as owners or possessors of the land, something that should be clear given our repeated inability to control our "possession." Our proper role has to do with the proper use of our rationality and safeguarding that reason from either ecological indifference—to biodiversity, the shared and interconnected systems of the planet, and so on—or impatience, which leads us off on all sorts of technological adventures without sufficient regard for the outcome. The key, argues Longbons, is to use our intelligence to mitigate the effects of our intelligence on other creatures and the wider world, a teleological use of our intelligence aimed at the mutual flourishing of humans and nature.

In "The Art of Creaturely Life," Norman Wirzba responds to Jean-Luc Nancy's question "who comes after the subject?" He does so by suggesting that *creatureliness is a more faithful and compelling rendition of human life than . . . modern characterizations of subjectivity that have often been uncritically absorbed by Christians." Creatureliness, says Wirzba, "goes to the heart of human identity and vocation, illuminating *who* we are, *where* we are, and *what we are to do.*"

He argues that the agrarian setting of the garden of Eden offers us key insights into the human condition, insights we have lost sight of: the way we are shaped by need and by dependence on the soil (*adamah*), on other creatures, and on the creator. As gardeners, we are reminded of the power of life and our interdependent connections with it, and we discover that soil, plants, animals, and elements are not mere "resources" for our use, "but are instead the fruit of a mysterious, fresh, enlivening power that transforms death into fertility and seed into fruit." In the face of this mystery we come to recognize our own limited power in the failures that threaten or accompany all our efforts, and grow in humility and appreciation of our creaturely nature as a result.

Wirzba notes that this reading of Genesis, which emphasizes creaturely need, finitude, and limit, parallels interpretations offered by Dietrich Bonhoeffer, in which limit is not seen as a deficiency but rather as a good through which we acknowledge that we live by grace rather than our own power. The transgressive act by which we deny these limits (and thus deny God) and seek to live autarchically can be seen not only in the Genesis account, but also in modern accounts of the subject critiqued by Nancy, Henry, and Marion. Wirzba concludes his reflection by outlining three marks of human propriety in light of our creatureliness: the acknowledgment of the goodness of limits and the refusal of the temptation to live *sicut deus* (as gods); the reception of life as a gracious gift; and, finally, the hospitable welcome of and service to others.

In "Face of Nature, Gift of Creation" Bruce Foltz points out the inadequacy of words like "environment" and "nature" for capturing what Hölderlin calls elsewhere "the wonderfully all-present." "Nature" and "environment" are shallow and inadequate words that immediately entangle us in problematic dualisms and scientific presuppositions. Heidegger suggested that *physis* is a more appropriate *Leitwort* (key word), one that facilitates the reenchantment of the world and, indeed, an awakening of the holy. Nietzsche, as well as recent interpreters such as Dreyfus and Kelly, has responded by suggesting we adopt a "new polytheism" that is attuned to the sacred meanings in the world. But is *physis*, as the Greeks used it, actually the word that gets us to the "holy" in the vibrant, dynamic, eventful way in which Heidegger uses the term? Foltz suggests that a better source, a better *Grundwort*, for what Heidegger hopes to capture by *physis* is *ktisis*, a word that is translated, though perhaps not very well, as *creatio* or "creation." Foltz sees in *ktisis* the "ever fragile, perpetually enacted emergence and renewal and sustenance of heaven and earth from the unthinkable nothingness *from* which they are ever brought forth and *to* which they always incline"; it is this dynamism that Heidegger sought to capture with the less appropriate *physis*.

When *creatio* is understood as "first efficient cause"—as opposed to *ktisis* as the performative act of "let there be" initiated by God alone—the path to the modern, disenchanted concept of nature is laid out, as is Heidegger's misunderstanding of *physis* vis-à-vis Christianity. *Ktisis* views creation as an icon, an opening to the transcendent creator; it is aesthetic and intuitive, as well as relational and personal; in it we see through the image to the original. Contrast this with the reified, naturalistic, and ultimately idolatrous concept of the *kosmos* as self-contained substance, and "nature" as something *vorhanden* though which we can infer some other absent, causal substance. However, to see the world as *ktisis*, we must undergo a spiritual recovery, a kind of purification of the soul that allows us to see anew what is, in fact, already there. The experience of *ktisis* affirms the materiality of the world: the view that the Incarnation is a fulfillment of creation rather than an anomalous event, which demands a soul emptied of selfishness and desire through *askesis*, and recognizes the difference between the divine essence (*ousia*) and the divine energy (*energeiai*) such that the experience does not devolve into a kind of pantheism.

In her essay "Creativity as Call to Care for Creation?" Christina M. Gschwandtner analyzes similarities between "cosmic liturgies" described by, respectively, Jean-Louis Chrétien and various Eastern Orthodox theologians. The Eastern Orthodox concept of human priesthood on behalf of creation—as opposed to the Western idea of human kingship or stewardship—appears to offer us a more ecologically sound interpretation of the human claim to be made *imago dei*. For example, Johan Zizioulas argues that the *imago dei* is an essentially *creative* characteristic, a calling that establishes our uniqueness. It means that humans have a "freedom in otherness" and a "drive toward love and creativity" that is manifested in the ability to "create alternative worlds" and "project themselves into a work of art and portray the other." Unlike other beings, we are called to the priestly act of creation and loving care for it: "The human being is called to bring the rest of creation into communion with God, so that the hypostasis of every creature might be saved" and its eschatological immortality assured. However, Gschwandtner sees real problems in these, and other, Orthodox approaches. Specifically, there is an implication that nonhuman creatures have no direct relationship to God, but rather require human mediation, a mediation that too often assumes forms of agricultural labor or domestication (as in Bulgakov). But how does domestication "redeem" nonhuman creatures? Moreover, certain thinkers develop human priesthood in a manner that suggests we are, in fact, "divine *by nature*," denigrating or erasing our very creatureliness.

Chrétien argues that "the human voice can become a sheltering ark for the rest of creation," summoning and responding to other creatures. Beauty orients us toward God and to the specifically human task: "by responding to the beauty of the world, we speak on behalf of the world and give speech to the world." In responding to this call, we become ourselves. At first blush, this seems similar to the creative and poetic human mission found in Zizioulas's work: both stress the call to creativity, and both express the response in terms of "lifting up" creation. However, Gschwandtner argues that while Zizioulas is much more explicit in his call to ecological con-cern—which, in his emphasis on creation, Chrétien does not specifi-cally address—it is nevertheless the case that Chrétien avoids many of the problems identified in Zizioulas and other Orthodox thinkers. In Chrétien's view, "the world as a whole offers itself to God," and therefore other creatures are also called and can also call us as part of a "larger chorus" in which we participate. Chrétien does not lose sight of our creatureliness—indeed, he stresses it—and his acknowl-edgment of vulnerability that prevents his work from casting human responsibility in terms of an all-too-often destructive superiority.

In "Creature Discomforts," Jeffrey Hanson argues that, contrary to other interpretations, the trope of creation *ex nihilo* plays, with some minor changes and development, a significant role throughout Levinas's writing rather than being abandoned after *Totality and In-finity*. Levinas uses the trope of creation *ex nihilo* in a primarily eth-ical, rather than ontological, sense: first, to underscore the absolute and immemorial character of the ethical demand to which we are called by the Other and, second, to stress that the singularity of the ethical subject cannot be absorbed in any totality: "only creation es-tablishes a truly separate subject." The infinite is not an expression of totality, but that which surpasses any totality, and this remains true even when it accommodates itself to a separated subject.

Paternity—the model of the child who both issues from the father in fecundity and who is nevertheless other to the father—is a way of expressing multiplicity without totality. It represents creation *ex nihilo*, in which the created being both issues from the Father and is absolutely other than the Father.

The main development of creation *ex nihilo* over the course of Levinas's writing is the degree of passivity necessary for a creature to be a creature, culminating in *Otherwise than Being* with the claim that the creature is more passive than passivity itself. The problem is that this evolution presents challenges for Levinas's ac-count: the more it emphasizes the meaning of ethical singularity the more it misses a convincing account of how ethical singularities remain a nontotalizing multiplicity: "the more Levinas insists that the ethical subject has nothing of her own apart from and prior to the bestowal of her subjectivity upon her by the Other, the more he requires but does not in fact provide something like the family, a community of singulars that stand in meaningful ethical relationship to one another without totalization." Hanson argues that Levinas's attempt to address this issue—the account of justice in *Otherwise than Being*—ultimately fails, and that a reappraisal of the role of the family, as begun in *Totality and Infinity*, is ultimately a better way to reconcile these two concerns.

In "Reflections from Thoreau's Concord," Edward F. Mooney turns to the concrete experience of living itself in order to reflect on "creation, on being a creature among others in creation, and on the role of creativity as a vector animating creatures and creation." How, in reading Thoreau's words, do we think with him and converse with him? When we follow him through the places he inhabited and visited—Concord, Walden, Ktaadin—to what extent do we too become residents or travelers in these locales? If, as Emerson says, "we are place," the distinctions inner, private, and hidden begin to break down, for in a sense we are then always "exterior." This suggestion might be developed to think Thoreau, as well as any other being, *is* creation. Here the "I" melts away as a superfluous distinction.

But if Thoreau is a philosopher, it is a philosophy of a distinct sort: one that believes wisdom is not mere knowledge, but rather a sort of living beyond accumulated knowledge, and one for which the goal is "sympathy with intelligence." Here sympathy is sympathy with and for the wonders of creation, sympathy that embraces "immersion, feeling, and tactile-intimacy" and that recognizes that "the life of the wild rouses the wild life in us." It searches for things to love, offers faith and hope for the future, and establishes "gratitude, remembrance, and . . . cherishing life" as core components of the exuberant, participatory philosophy for which Thoreau advocates. True, we remain wounded animals (just as we are rational animals, political animals, and so forth), caught in the circle of life, death, and rebirth; but for Thoreau this state is bound up with grief *and* celebration, in which art—including writing—rescues the world from "abandonment" and "meliorates the wounds and trauma we endure."

Janet Martin Soskice reflects on the topic of *creatio ex nihilo* in her essay "Creation and the Glory of Creatures," arguing that we must be careful regarding the easy slippage between "creation" and "nature," since, speaking biblically, "creation" refers to the activity of God and the outcome of that activity rather than to the narrower sense of material nature. "Creation" implies a creator, while "nature" does not. And the doctrine of *creatio ex nihilo* is the teaching that "God has created all that is, including matter, space, and time, and does so freely out of love." Here God's created power is always linked to salvation, resurrection, and God's power to recreate.

Although the nature of biblical creation is still a matter of intense debate—as is whether creation took place *ex nihilo* or *ex hyles*, via the manipulation of some primordial element or matter (as possibility indicated by the Hebrew *tohu wa-bohu*)—what seems uncontestable is that Jews and early Christians thought about creation in terms of the freedom, power, and goodness of God.

From where, then, did the notion of creation *ex nihilo* come? It is foreign, for example, to both Plato and Aristotle, and it did not rise to prominence until second-century debates with Gnosticism. Soskice argues that the notion is not Hellenistic or philosophical in origin, but that it arose as a biblically inspired piece of metaphysics in the thinking of Hellenistic Jews. Despite widespread belief that he is dismissive of *creatio ex nihilo*, in the writing of Philo of Alexandria we can find "all the principles central to the doctrine of divine transcendence that are enshrined" in that teaching. Philo, a faithful Jew who is primarily a scriptural exegete rather than a philosopher, brings to middle-Platonism his piety and reverence for the transcendence and sovereignty of God. In turn, the doctrine of creation *ex nihilo* and the divine transcendence it implies lie at the root of debates about a theology of divine names, or of apophatic theology, even as found in thinkers not normally associated with those traditions, like St. Augustine. Thus we ought not complain that "the wholesome streams of Judaism and Christianity" were "contaminated by Hellenistic philosophy"; rather, we should recognize the Christianization of Hellenism was perhaps more significant here than the hellenization of Christianity.

In "Care of the Soil, Care of the Self," T. Wilson Dickinson interrogates our familiar places and habitual practices in order to see anew the ways of being of obscure places, ecological connectivity, and the relationship between creativity and creation. He suggests that the care for creation may be connected to the formation of character, and vice versa, in surprising ways. The project of analyzing connections between social, religious, and ecological spheres is well-established: the Hebrew scriptures, which are grounded in an agrarian outlook, emphasize our duty to care for the land; and the Hellenic techniques and practices of caring for the self were adopted by Christian philosophers. However, contemporary suburban life—from strategic or systemic leveling of nature in service of growth down to quite commonplace fixation on things like lawns—seems to cultivate ignorance or even contempt of place. Lawns generally require the application of fertilizers, pesticides, herbicides, and irrigation systems heedless of the intricacies and nuances of place. This process cultivates hubris and conformity and undermines human autonomy and creativity. Theologically it is "the expression and catechesis of a realized eschatology that identifies the Kingdom of God with the Manifest Destiny of consumerism." However, if the lawn is one type of school forming one sort of character, Dickinson argues that the "garden school" offers a different sort of education, one that cultivates attention, appreciation of complexity and mystery, patience, humility, invention, and similar qualities. Another sort of school can be found in the wild chaos of the field, in which one can come to appreciate the beauty in strangeness and otherness.

Finally, in "Dream Writing Beyond a Wounded World," Susan Pyke argues that literature can shift people's perceptions of the nonhuman world and, in so doing, play a central role in forming a new ecological consciousness. In particular, when they are not written from the familiar anthropo-theological perspective, such works have the power to illuminate the intimate connections between country and life and the ways in which communion and communication with it can shape our choices and move us to action. Canadian poet Anne Carson's poem "The Glass Essay" is an example of such ecological "dream writing." This poem offers us a disruptive rereading of Emily Brontë's *Wuthering Heights*, one that "accepts the sentience of the non-human" and connects readers to matters outside the sphere of human significance and concern. This alternative vision of implacement posits a "porous flow between habituation and self" in which humans are "part of a community of matter that includes flora, fauna, land, sea and air." As the "country" of the Yorkshire moors speaks to Brontë's Cathy, a parallel exchange takes place between the Canadian moor and Carson's anonymous speaker; and in reading of these exchanges, we too may become sensitive and experience "topographic invocations" that call us to an appreciation of the interplay between place and person, matter and spirit, and that summon us to a "posthuman love."

Creation, Creativity, and Creatureliness: The Wisdom of Finite Existence

Rowan Williams

Bishop Kallistos has already mentioned the significance in the twentieth century of Father Sergei Bulgakov as one who elaborated a comprehensive theory of the wisdom of God as the key to understanding a whole range of theological issues, and indeed issues more than just theological. (If any issue is "really more than theological"; perhaps I should say less than theological!)

But Father Bulgakov's thought has often seemed impenetrable to the casual Western reader, or even the not-so-casual Western reader, and the not-so-casual Eastern reader, as well. It has seemed to be a piece of metaphysical elaboration without immediate relation to the heart of the gospel. It was a very distinguished and saintly Russian cleric who said that his first reaction on reading Father Bulgakov was, "they have taken away my Lord and I do not know where they have laid him."

I hope in what I say, with due respect to that particularly saintly Russian, to suggest that this is an erroneous apprehension of Father Bulgakov and that properly understood his account of holy wisdom opens up a way of thinking about—quite simply—life in Christ, life in the body of Christ, life in Christ's world, which gives us some remarkably challenging and resourceful ideas for coping with that bizarre environment in which we live called modernity.

23

But a word first of background and explanation about Father Bulgakov's theories: He stands in a long tradition of Russian speculation about the wisdom of God, a tradition whose greatest nineteenth-century advocate is Vladimir Soloviev, for whom holy wisdom, Sophia, was a visionary presence, a kind of eternal feminine, permeating the world and accessible to human vision at certain times. Soloviev wrote about his own encounters with Sophia as kind of person. And at the very beginning of the twentieth century, the great Father Pavel Florensky, perhaps the most formidable mind of his generation in Russia in almost any sphere, elaborated this further in a work under the title *The Pillar and Foundation of the Truth*.[1] He drew on liturgical and iconographical themes to sketch a vision of the universe permeated by wisdom, which has deep roots in certain aspects of orthodox theology, although it goes a good way beyond it in many other respects. But it also has roots in the neo-Platonic and esoteric speculations of the seventeenth and eighteenth centuries in Europe and reaches out its hands to a whole variety of disciplines and perspectives in the cultural and religious world.

Father Pavel was one of those who helped to draw the young Bulgakov, professional economist, Marxist, and aspirant politician, into the life of the church. And it's not surprising that Father Bulgakov's early works are very deeply marked by Florensky's influence. But as Bulgakov himself matures as a thinker, other things begin to happen. And what I see most clearly in the evolution of Bulgakov's work is a drive toward a more comprehensively biblical and liturgical setting for this language of holy wisdom, until in the great trilogy of his maturity we see a Christian systematic theology oriented around the theme of holy wisdom, very largely purged of some of the ambivalent and perhaps unduly complex or even syncretistic aspects that you might find in Soloviev and in others.

But to try and crystallize exactly what is the heart of this vision of the world you would have to say something like this: Holy wisdom, rather like the principle of "logos" in the philosophy of Philo of Alexandria at the beginning of the Christian era, is what you might call the area of overlap between divine and created life. If you think of intersecting circles, holy wisdom is that point at which the realities meet, overlap, intersect, interpenetrate. Wisdom, in Bulgakov's mature thought, is not some kind of extra person in God; he toys with that language as did Florensky, then he sensibly goes somewhere else with it. It is not an extra person but a quality of the divine life that, in an early work, Bulgakov refers to as the love of loving. It is God's own being reflecting lovingly upon its own loveable-ness, and in that reflection and relation opening itself out to the sharing of love beyond the divine being. Thus there is a quality of loving wisdom in the very heart of God's life; it is that quality that is then shaped in the heart of the world. The mode, the rhythm, call it what you will, of divine life becomes the wellspring, the central energy of created life also.

So you can talk about "uncreated" and "created" wisdom, for example, as Bulgakov often does. Wisdom in the life of God is that in which God thinks of all the things that he will love. In loving his own productive, generative, generous love, God loves all those ways in which that love can be realized in creation. From all eternity God is loving the possibilities of his own creativity, so all things are held together in that loving wisdom. And when the world is created, it is created in such a way that those eternal objects of God's loving wisdom become actualities—interacting with one another, relating to God in the finite realm. So in creation, wisdom means the pattern of hidden harmony, interaction, interpenetration between the life of all lives in the world.

And one more word about these basic issues: it's out of this that there arises Bulgakov's distinctive contribution to the understanding of Jesus Christ, in whom uncreated and created wisdom come together. In Jesus Christ there is the perfection of created life, in him all the harmonies of the universe converge, "in him all things cohere," in St. Paul's words, and this is possible because he is also and eternally the Word of God in whom the intelligence and love of the Father are realized. But more of that in a moment.

That, in somewhat abstract terms, is an outline of sophiology. What it doesn't tell us is how it makes a difference to our self-understanding and our understanding of God in a huge range of different contexts. And that is why I would like to argue that Bulgakov, perhaps more than any other of the great theologians of holy wisdom in the Russian tradition, is crystallizing something crucial to faith, not simply speculating.

He is challenging us to find a way of understanding two related things that are indeed central to the Christian enterprise. He is challenging us first to understand how God and the world are related in a continuity that is also an unfathomable difference. And Bulgakov, like many Christian theologians, would say that the greatest errors of Christian theology come when Christian theologians become preoccupied with either the continuity or the difference in such a way that they cannot understand how grace works. But the second issue, related to that, is more specific to our humanity. So it is not only about understanding God and the world in continuity and difference, it's also about understanding ourselves as being in the image of God—understanding the particular, the unique calling of human beings in the universe like this, made by a God like this.

Some of this may be illuminated if we wind back our theological history a little bit to the fourth century, in particular to a remark of St. Athanasius in one of his treatises, "Against the Arians,"[2] where he says (almost in passing) that if God were not eternally the generator of the Son, the Word, we couldn't understand how God could be creator. It is an observation of huge importance that lies at the root of the tradition that I'm trying to reflect on. Creation is not necessary to God; God can get along perfectly well being God without creation. But, to put it in a slightly trivializing way, it is not at all surprising that God is the creator, that God is eternally one who generates what is other, who eternally makes different his own life in the outpouring and exchange of the life of Father, Son, and Holy Spirit—that tells us that in the heart of God there is what you might call the energy of difference, an outpouring of life into otherness. If God were not like that we could not understand creation; or rather, we might understand creation in a completely misleading way. We might understand it as the product of an arbitrary divine will. We might understand in the deist way in which many did in the eighteenth century—a God whose power is expressed in making a universe in which he has no living investment. But creation as the carrier of wisdom, creation as a "participant" in some sense in the divine life, makes sense when we understand our God as a God who makes himself other, who is in a relation of loving difference. So creation is, you might say, the sort of thing God does. Creation is a free outpouring of what God is; but that free outpouring makes sense because God simply is the Father of the Son, the breather of the Spirit from all eternity. What is natural to God as God in being Trinity is then freely and willingly shared in the act of creation.

So as we grasp the unity in difference of the Father and the Son, something comes into focus for us about the nature of creation itself. I turn here to one or two remarks from one of the works of Bulgakov's maturity, his great work, *The Lamb of God: On the Divine Humanity*.[3] The trinitarian God who is love is in his wisdom creator, in virtue of that very fact about his nature. In no sense does he become so at a given moment in time when the creation of the world begins. The world's creation has a beginning from the world's point of view, not from God's. What in God is eternal is manifest for creation in time. In this respect a certain translation from one language to another is indispensable. Creation translates into time and limit and history the eternal fact of God.

Now in that work on the Lamb of God, Bulgakov offers some very searching reflections indeed about the nature of God's self-giving. And after a longish historical introduction on the doctrine of Christ, he begins his substantive and creative theological discussion by concentrating on this theme of the divine self-giving. Here Bulgakov speaks of *kenosis*, that fundamental theological concept, the self-emptying of God. Eternally, says Bulgakov, God the Father empties himself out in the generation of God the Son. Eternally God the Father decides to be only for the Son. And that eternal self-emptying in the generating of the Son again gives us a clue to the nature of creation. Before the Word of God empties himself to take on human flesh, the Trinity is involved in a self-emptying act in shaping the world. That the world should be is for God to withdraw but not to be absent. It is for God to let be a world with its own freedom, its own integrity. The God who creates a world of freedom, a world that is itself, is a kenotic God, a self-giving, a self-emptying God whose being is for the other. And as we understand this in the eternal life of the Father, the Son, and the Spirit, we understand how it is in creation.

And what Bulgakov is hinting at in *Lamb of God*, what is developed in the long and very beautiful argument of that great work, is that the understanding of divine wisdom is inseparable from the understanding of divine self-giving. If we want to understand wisdom we have to understand self-giving, self-surrender. In theological shorthand, wisdom is kenotic; to live in wisdom is to live in and by this energy of dispossession and outpouring. And that's why wisdom for the mature Bulgakov is not any kind of quasi-personality but the tone, the quality, the mode of God's very being as shared, as outpoured and returned in love; the eternal active life that is God, which is the love of love; the love of what self-emptying makes possible, given, returned, circulating eternally between Father, Son, and Spirit. Creation can come to a full share within its finite terms in the life and the liberty of God because God lovingly "withdraws." Creation participates in God because God makes space for it. And here Bulgakov draws on the very rich heritage of Jewish as well as Christian mysticism—on divine withdrawal as one of the fundamental themes in Jewish mysticism of the late Middle Ages and the seventeenth and eighteenth centuries.

Creation then is to be understood as that which is other than God and yet in being other than God is exactly what God desires, because God desires to give and realize his love in what is other. That eternal pattern of the Father, the Son, and the Spirit is translated, to use Bulgakov's terms, into the relation of God and creation. And because that creative act is essentially an act of self-forgetting, self-giving, self-sharing, creation becomes itself when it lives into the reality of self-giving, self-sharing.

And so I make my first link from creation to creativity. The human calling to share the love and the liberty of God has to be in this perspective a calling to "let be." The paradox of real human creativity is that it is not the flexing of our human, our created will, the flexing of our muscles, the imposing of order, the dredging of up of something new out of the depths of our interiority; our creativity is most fully and freely expressed as humans when we, as artists, stand back and let be. How many artists have written about this aspect of the creative enterprise? That it is so far from being an exercise of will, much more an exercise of the most extraordinarily concentrated suspension of will, so that something is allowed to happen. The really creative work is a happening; it is the depth of the world occurring where the artist is because the artist has somehow exercised that asceticism of setting aside preferences and purposes and all the rest of it, so that something occurs. Many years ago I used to meet regularly with a rather formidable Canadian sculptor who wanted to talk about what he was doing. And he would say passionately again and again, "will has nothing to do with it." I must let it happen. Because the paradox is that "letting it happen" for any artist is a hugely laborious business; it's not a recipe for sitting back. But that is simply to say the creative artist is doing a kenotic job; allowing the rhythm of the deepest reality to become transparent in your acts, your imagination, requires a real discipline of self-forgetting. And of course that's why in Bulgakov's own writing, particularly in some of his earlier writing, there's a very deep connection between the creativity, the human creativity that goes into art, and the human creativity that goes into the building of society. Creativity and justice belong together for Bulgakov.

In some of his own writing, Bulgakov speaks in an early work about art in these terms:

Does art not save, does it not alleviate the anguish of existence on this earth? Does it do no more than a muse? Must it and should it be no more than an entertainment that has no power to change anything? It is possible for products of the arts to be admired and loved in their own right, but this only has the effect of making the chains' vile existence more palpable, and it is impossible to love these products with a living human affection.

So every creative act strives to attain an absolute status; it longs to create a world of beauty to triumph over chaos and convert it to order. But what does it actually say or convert? The artist, even if the greatest artistic attainments are granted him, has all the greater a sense of unsatisfactoriness or frustration as a creative personality to endure. Creativity is a stony path where the weight of the cross is laid on the shoulders of Simon of Cyrene whether he wills it or no.

Bulgakov is saying there, and says it in many other places, that art can never be content with producing a beauty that is just there to be looked at. Art wants the world to change, and in its uncovering of the "sophianic," the wisdom-directed depths of the world, it does indeed make it possible for the world to change. But it does so fully only alongside those other activities of creativity in our human relationships and our society that bring to light the moral and spiritual depths of wisdom. So beauty is not just about art, and if Dostoyevsky said, "Beauty will save the world,"—a favorite quotation of Bulgakov's—he didn't just mean that it was nicer to have nice things around. There's a real transformative and revolutionary element here about that great act of self-dispossession that in the artist or in the fighter for justice allows truth, reality to come to light.

This interests me very deeply, not least because I've been working recently on some of the ideas of Bulgakov's contemporary Jacques Maritain, the French Catholic philosopher who in the 1920s was writing about art and faith in some very searching ways. One of Maritain's greatest themes was that the artist seeks the good of what is made—not the good of himself or herself, but the good of what is made. The artist seeks to make something that is itself, not just the conversion of something "in here" to something "out there," but something that has a life, an integrity of its own. And that is the challenge, the agony at times, of real art, not the passion for self-expression, but something completely other than that, a passion that there be something that is itself. And Maritain has a series of reflections on what this implies for the understanding of art when it truly is about letting something other be. I think there are many links to be explored between what Maritain has to say about art and what Bulgakov says about it, noting also that both of them share the same conviction that art cannot be separated from the quest for justice, that the passion, the "sophianic," wisdom-rooted passion, to undercover the truth of the world leads you to seek justice and reconciliation as part of the search for beauty.

Creativity in the created world becomes then a mirror of God's nature. Not in a general, rather vague way (God is the Creator, and human beings are quite creative, too), but because God's self-forgetting in creation is the model for our own discovery of reality in self-loss and self-denial—the denial of the selfish will in the artist's work, the denial of a crude individualism in the social realm. Labor and creativity in human relations are as much an act of holy wisdom as is the artistic enterprise and as much a kenotic matter, a "being for the other." Bulgakov was thinking about this very early on, as early as 1912 when he published his book on the philosophy of economy.[4] Any student of economics picking up that book expecting to find detailed discussions of markets and surplus values would have had a nasty shock, because it's mostly about God and ethics and justice and things like that—which economists are (strangely) not interested in a lot of the time.

But of course the implication of that for our situation in the modern world is quite challenging. It means that campaigning about debt or fair trade is creative, it is an exercise of what our humanity is called to, it is the kenotic and sophianic search for a justice that is beautiful, a justice that uncovers what the world fundamentally is: a world of interdependence and interaction, a world in which self-forgetting brings joy, common, shared joy. And very often Christians have somehow failed to get across any idea that ethics, whether individual ethics or social ethics, is about joy. Those two words—ethics and joy—you may not habitually associate; but that is a theological failure, because the search for justice is very profoundly a journey into joy. If it's true that this is what the world is, if it's true that the nature of our participation in the life of God is a participation in God's self-forgetting bliss, then our work for a society in which people have the freedom and the dignity to give themselves to each other in love, is as creative as any other act we undertake.

Bulgakov doesn't elaborate on this at great length, but there's actually quite a lot of work to be done on what ethics would look like in the context of a doctrine of holy wisdom like this. And those great thinkers who in the last fifteen or so years have begun to shift Christian ethics toward a different, church-related model are moving in precisely this direction. Whether it's Alistair MacIntyre, Stanley Hauerwas, John Milbank, or Vigen Guroian (himself an Armenian Christian)—all of these writers in their different ways seem to be to be saying that ethics is to be understood only in the context of the body of Christ. We are not talking about duties and rules, nor are we talking about the freedom of self-expression. When we talk about ethics we are talking about the realization of wisdom—not human prudence, but the self-sacrificing wisdom of God that is the heart of all reality and the center of all life.

If we were then to think of ethics in that context we would perhaps begin to see more clearly the sense in which we can speak of the church as a place where creation is itself. Often that seems rather counterintuitive; to say the church is the future of creation, the church is what humanity is about, the church is the promise of fulfillment—matches only a little unevenly with the reality of many Christian communities that we know and love. And yet the church as the place where matter is transformed into divine gift, the church as a place where persons exist and exist only in communion, in mutuality, in an utter being there for each other—in that sense what can we say but that the church is indeed the future of creation? The church is where creation is itself, the church is where wisdom is realized.

The day-to-day realities of the church may take some time to catch up with this vision. It has always been so; we shouldn't expect it to be otherwise. But what else is going on in the church but wisdom? Here the things of the world, the bread and the wine at the Lord's table, become unequivocally, unreservedly the vehicles of divine gift; the stuff of this world becomes soaked through with the self-giving of God because the bread and the wine of the Eucharist, given into the hands of Jesus, become imbued with his sophianic life, his life as the life of self-giving wisdom. Given into his hands and then received from his hands, they become the carriers of wisdom into our own lives. And as that life is given back to us in Christ, the life of the community becomes the life of wisdom realized, because in that moment when we are the body of Christ as we have received the body of Christ, each of us is uniquely and gloriously there for all others. The moment when we receive the elements of Holy Communion is the moment when we are endowed with the capacity for self-giving, being for the other in the way that only each one of us can be. That is why as we come back from the receiving of Holy Communion, we ought to be awed and amazed at the people next to us in church—because they are imbued with divine gift. And rather as in the conversation between St. Seraphim and Motovilov, each of them could say to us "Your face too is alight with the love of God." That's the sense in which the church is where wisdom is and where every church community worth the name is struggling, feebly and patchily and very unsuccessfully, to be a place where wisdom is visible, matter transformed into divine gift, persons existing in communion. That's the background against which we ought to be talking about ethics as ecclesial, ethics as a church matter, and ethics as a matter of joy.

So we come finally to the celebration of creatureliness. Arguably what is going on in the work of redemption is, as St. Irenaeus first put it, the reversal of Adam's mistake. Adam's resentment at not being God is transfigured by Christ into the free acceptance of not being God. That's what Philippians 2 is all about. The one who is in the form of God delights to be no longer in the form of God but in the form of a slave, and in that slave form of humanity, joining in our unfreedom, our suffering, our tensions and our struggles, the finite created form of humanity is glorified from within. Adam resents not being God and so Satan has leverage upon him: "You shall be as gods," says Satan to Adam, knowing that the essence of our fallenness is resentment at being creatures (just as the essence of the fall of Satan himself, in church tradition, is the refusal to worship). So Jesus, in not clinging to the form of God but accepting the humility of the incarnation and the death of the cross, restores the glory of creatureliness. The incarnation affirms that creation is good, not that it is nice or beautiful, but that it is good because it is in this relationship of loving dependence on the self-giving of God. And the mystery that we seek to understand when we think about redemption is that restoring of the glory of creatureliness can only be done by one who isn't simply a bit of creation—the Word in whom creation hangs together, in whom alone is that full freedom that can accept the otherness, the suffering, the death of the created order and fill it with life. "He who ascended, is it not he who also descended?" (Eph. 4:9).

So we in Christ rejoice at not being God. We ought to give thanks daily to God that we are not God and that God is God; we give thanks to God for God's great glory. And the secret is that only in that rejoicing that we are not God do we come to share the divine life in the way we are made to do—the paradox that only by our completely not wanting to be God can the divine life take root in us.

Discipleship in the body of Christ is in one sense simply a matter of constantly battling to be a creature, battling against all those instincts in us that make us want to be God or make us want to be what we think God is. There, of course, is the catch. And that's why discipleship challenges at every level those unrealities that distort humanity, that distort creatureliness. That's why discipleship challenges those enterprises in our world and our culture that feed the illusion that actually we could be God if we tried hard enough.

What are those things about? Well you may find them in the deep unease so many in our culture feel about ageing and dying. You find it in our denials of death. You find it in our passion for absolute security, our desire never to be at risk. You may find it in a defense program; you may find it in the technological exploitation of the environment. At level after level, our temptation is to deny that we are finite. And when I read, as sadly I sometimes do in discussions of our environmental crisis, that we can be confident technology will find a way, my blood runs cold, because I hear in that the refusal of real creatureliness. "These limits are temporary, our skills will find a way, we shall at some point be able to get to the stage where we are safe." And the gospel tells us you never on earth get to a place where you are safe; but you will get to a place where you are blissful and united with your Father in heaven. In the immortal words of C. S. Lewis, "he's not a tame lion, you know."

The outworking then of created wisdom, created Sophia, is this joyful embrace of being created, of not being God, the acceptance that we shall die, that we are fragile, that we are fallible. And it is "here on this lowly ground," in John Donne's phrase, that we come into contact with the transfiguring, transformative life of the eternal God. Creation, creativity, creatureliness—you see perhaps the connections that I'm trying to draw out between those three clusters of ideas. When we understand what it is for God to create and how that is rooted in his trinitarian being, characterized by holy wisdom, then we begin to understand how creativity works in our world. And as we understand how creativity works, how it is always bound up with love, with that bracketing of the self to be for the other, that will for the good of what is made and so on, then we see that our holiness is not the denial but the acceptance of being creatures— made possible in that great central mystery of the creator himself becoming a creature, uncreated love working through the created humanity that is Jesus of Nazareth so that created mortal life is touched and glorified.

The value and virtue of language about holy wisdom, the "sophiological" vision of Father Bulgakov and others, are that there are actually very few models or metaphors that really hold these three things together—creation, creativity, creatureliness. These desperately need to be held together for the integrity and transforming strength of our Christian language. Sophia, holy wisdom, becomes the line between God's nature and ours. It makes sense of the idea of universal "kenotic" energy, the universal urge in creation to that self-emptying love. It makes sense above all of Christ as the point where created and uncreated meet in such a way that everything is changed by the meeting. So, granted all the difficulties, granted all the challenges that someone like Father Bulgakov places before us in understanding a highly sophisticated and intensely convoluted metaphysical model, perhaps we should bring it back repeatedly to this simple center: Christian faith is about God and the world in continuity and difference; Christian faith is about the image of God in humanity and how it is restored and reenergized. To speak of holy wisdom in this context is to speak about all these things and to give ourselves not simply an intellectual agenda but one that is spiritual and creative, and public and political too—all inseparably linked together as we seek to work with and for God in a world that God wills to share his joy, that God has made to share his joy.

Rowan Williams
and Ecological Rationality

Jarrod Longbons

I

Ecological anxieties loom large in the popular mythos. Films such as *Avatar*, television programs like *Life After People*, "Climate Gate," and countless *feuilletons* reveal widespread concern about the health of our planet. Amidst the wilderness of messages is a voice crying out like Nietzsche's "mad man." His name: Slavoj Žižek. But whereas the "mad man" announces the death of God, and too early for his time to hear it, Žižek announces the death of ecology, claiming that ecological obsession is ideological. This ideology says that the ecological crisis is the result of human activity, because modern humanity is itself "becoming a geological agent."[1] This ideology comforts, because if humans cause the problem, then "everything depends on us . . . we can save ourselves by changing our lives."[2] For Žižek, ecology is the new opium of the masses. (One wonders, is his message too early for his time to hear it?)

Whereas Žižek names our ecological concerns as ideological, former Archbishop of Canterbury Rowan Williams views the ecological burden as an opportunity that causes society to rethink life with a necessary ecological rationality and that can help us rediscover some of the implications of the Christian doctrine of creation. Williams joins Žižek in the wilderness to voice a Christian response to the growing concern of ecological health. This essay will present these contrasting viewpoints, with Williams explaining and applying creation theology and Žižek representing a growing position of belief within certain strands of continental philosophy and ecological thought.

Žižek's diagnosis is not unreasoned. He begins with the idea that nature is not a harmonious, neutral state. Rather, in his typical reductive stance, he asserts that the environment is nothing more than one series of catastrophes after another.[3] And true enough, but one must also account for the life that emerges from the death and destruction of other beings in order to truly account for totality of life in our universe. Yet he goes further by asserting that "Nature doesn't exist: 'nature' qua the domain of balanced reproduction, of organic deployment into which humanity intervenes with its *hubris*, brutally throwing off the rails its circular motion, is man's fantasy."[4] For Žižek, we must dispose of this delusion, for it creates unrealistic expectations for human societies. Rather than an entity "over there," apart from humanity, nature is coextensive with humanity, and societies, he suggests, do not merely act upon a passive world. Žižek adds that "nature on Earth has already 'adapted' to human intervention to such an extent—human 'pollution' being already deeply implicated in the shaky, fragile equilibrium of 'natural' reproduction on Earth—that its cessation would cause a cataclysmic imbalance."[5] Has the earth really adapted to account for all human activity? To a degree, perhaps, but such speculation ends in *aporia* for humans regarding Earth's threshold for abuse. Further, should this be our only concern? Genuine ecological concern should extend beyond discovering what lifestyles societies can "sustain" without killing everything; rather, it should extend to, as the Christian tradition would have it, the flourishing of all things. Or as Aldo Leopold famously quipped (even though he was critical of the Abrahamic tradition as it relates to ecology), "a thing is right when it tends to preserve the integrity, stability, and beauty of the biotic community. It is wrong when it tends otherwise."[6]

The ideology that makes Žižek's diagnosis possible is nature-culture dualism, an ontology that posits the world of nonhumans (nature) and the world of humans (culture) living side by side in mutual exclusivity. For Hans Jonas this ontology disenchanted both the world and human relationships with it. "Dualism continued to drain the spiritual elements off the physical realm—until, when its tide at last receded, it left in the wake a world strangely denuded of such arresting attributes."[7] One result of this expensive ontology is that it provided much of the ethos of much modern scientific industrialization.[8] Indeed, for industrialized societies the nonhuman world's *raison d'être* is as a resource for human happiness and convenience. This dualistic stance derives, in part, from one of modernity's chief proponents, René Descartes, who claimed, ontologically, that humans are the "lords and possessors of nature."[9] And Descartes was not alone; even Francis Bacon, another catalyst of modernity, sought, via the laboratory, to investigate nature in order to learn how humans can master it.[10] We live in the wake of this technological ontology, and so ecology today often appears as a subversive worldview. This is also why we are witnessing "a recurring criticism," at least within anthropology, but also within other spheres "that the nature-society dichotomy hinders true ecological understanding,"[11] because it denaturalizes and thus disconnects society from the rest of the world. Simply put, this dualism denies that culture is natural or that it took time to evolve within history; rather, dualists propose a culture that is *ahistorical* and transcendent over the nonhuman world (many ecologists, in opposition, offer transcendence to nature over culture; both moves miss the mark, however, because they cannot account for either sphere without reducing the other).

As should be clear, dualism is the prevailing ontology of modernity, but Žižek's ontology is more postmodern, where one realm collapses into another or, as Žižek would have it, culture collapses into nature. For him, culture is folded into nature, which is why he argues that true ecology must embrace human trash. Society must accept its ecological footprint because it is as natural as anything else.[12]

Is Žižek's answer realistic? Certainly, humans are "natural," but what about human *poesis*? Popularly, human making—technology, art, customs, laws, religion, etc.—is considered unnatural because it appears different from the way humans first encounter the world.

But if, as I am arguing that humans are natural, wouldn't it be wrong to think that human work is of another kind? This question reveals an impasse created by nature-culture dualism, and this essay seeks to overcome it by rethinking the world as creation—an idea that brings the subject (culture) and object (nature) together under one heading without denying difference. But we must also note that human making is, as a result of the subjective mind, in some sense different from "nature" because humans reassign existent matter for alternative purposes.[13] The real problem arises with the question of whether or not human making offends the "natural rhythms of the world."[14] Or, as Bulgakov would argue, the pathology of modern technology, driven by our dualism, is that it wishes to make the human divine, which is a perverse simulacrum of the divine-human promoted by Eastern Orthodox Christology.[15] When human activity is deleterious to the world's natural rhythms, or when it seeks to make humans divine, it becomes parasitic.[16] But when following these rhythms, thus living faithfully within creaturely limits, human activities support all life and reveal an ontology of interdependence. Beyond "saving the planet," helpful human activity extends to flourishing of the entire biosphere. Thus this essay will consider the world as "creation," not merely *natura pura*, and will consider humans under the category of "creatureliness," not merely autonomous aliens on an indifferent planet.

With Žižek in the aforementioned wilderness is also the voice of Rowan Williams. For Williams, humans are natural but not homogenous with nonhumans. Humans differ from nonhumans most notably because humans are intelligent, boasting rationality beyond anything found in the realm of nonhumans. Williams indicts this intelligence because it enables advanced human *poesis* that often harms other species. For example, advances such as factory farming, nuclear technology, and the automobile have all eroded different ecosystems via their propensity for waste. Let's take factory farming as a concrete example. In the past, the farm was a cyclical milieu. Grass was grown to fuel cattle. Cattle provided the fuel to till the soil. Their waste fueled the soil, and finally when they died their bodies became fuel for the soil. Waste, before advanced agriscience, was not merely waste like the emissions from modern tractors, which is just one way that the modern farm harms an ecosystem.[17] There are also the examples of soil erosion and monoculture and thus the denial of biodiversity, carbon emissions, etc. But Williams does not denounce human rationality à la Ani DiFranco, encouraging humans to be "just like other animals."[18] Rather, he argues that "right" rationality is the key to understanding the meaning of human relationship with nonhumans: reciprocity. Right rationality, or ecological rationality, leads humans to live interconnectively with nonhumans, "to understand that we and our environment are alike in the hands of God, so that neither can be possessed absolutely, is to see that the mysteriousness of the interior life of another person and the uncontrollable difference and resistance of the material world are connected."[19] This kind of thinking outflanks nature-culture dualism better than postmodern blending, for it reveals human *relation to* and *responsibility for* nature, despite the obvious differences between these two classes of creatures.[20] To put it another way, creation avoids the hard segregation and alienation of dualism as well as avoiding monistic devaluing of difference. Creation promotes a paradoxical view of unity within diversity of being.

II

In developed cultures, the separation of nature from God and man leads to cultural exploitation of nature. Moreover, Williams finds this dualism promoted by many strands of modern scriptural exegesis.[21] An example of this is millennialist theology that prioritizes the salvation of the immaterial human soul; in such thought, the current world will end in destruction, so why bother trying to save it? But such readings offer only a shallow understanding of the fullness of the redemptive drama woven throughout scripture from beginning to end.

Robust readings of scripture, on the other hand, reveal that God extends redemption to the land as well as human communities. In fact, the biblical narrative describes a God that is as interested in the health of the land as he is with the morality of the covenantal people. Actually, the moral life of the ancient Hebrew included his treatment of the land—that is to say that the covenant is made up of three characters in the drama, not two: God, the People, and the Land.[22] Consider the warning of Leviticus 26:

> Then the land will enjoy its Sabbath years all the time that it lies desolate and you are in the country of your enemies; then the land will rest and enjoy its Sabbaths. All the time that it lies desolate, the land will have the rest it did not have during the Sabbaths you lived in it.[23]

Inattentiveness toward the land's well-being brought God's judgment. For while God gifts the land to his people, they never fully possess it. It remains God's. Williams says of the Old Testament community, "the people and the land alike belong to God—so that 'ownership' of a person within God's chosen community is anomalous in a similar way to ownership of the land."[24] Though possession is denied morally through biblical law, we cannot deny it practically, either. Consider the invasive Asian carp of the Illinois waterways. In the 1960s "the fish were originally brought to the country as an environmentally safe alternative to chemical treatment"[25] in a "natural" attempt to clean catfish ponds and increase the fish market supply. Eventually, flooding allowed these fish entry into the Mississippi River; their populations grew and currently threaten to spread into the Great Lakes. Since their diets and populations cannot be sustained in this habitat, the plan to help Illinois markets is now causing harm. In short, no amount of human sophistication controls creation; at most, society merely affects it,[26] but without the competence for actual possession. Attempting possession often results in unforeseen, detrimental circumstances. The ontology of creation, however, as opposed to *natura pura*, is based on reciprocal relationships within creation and a relationship with the creator via a "nonreciprocal relation of dependence."[27]

True rationality leads us to this same conclusion. So Williams argues, "if it is more than us and our relation with it, it can survive us; we are dispensable. But the earth remains the Lord's."[28] Ecological rationality corresponds with theology: humans, like nonhumans, are contingent beings who participate in a given being and constantly need one another for survival. The logic of ecological rationality (and Christianity) is exceedingly more humble than that of modern industrialism, which overextends human efficaciousness; it naively thinks society can control nature, can account for all ecological burdens, and therefore is poised to alleviate them. This is Žižek's insight. But Williams argues that humans are not the solution because the earth is not ours to save. All humanity can do is exercise its God-mandated vocation, rediscovered by renewing ecological rationality. Furthermore, this vocation must be enacted even if it does not yield desired results. The point is that value is inherent in the whole biosphere because it is imparted by the first gift—creation—so our ethical stance should exceed mere pragmatic goals.

To repeat, Christianity confesses that the whole *cosmos* is the subject of the redemptive drama of God, the creator who formed all things *ex nihilo*, good, and for his own purpose. Williams summarizes Ellen Davis's work: "the language of redemption applies both to the land and to the people; both are in God's hands, and thus the people called to imitate the holiness of God will be seeking to save both persons and property from being alienated forever from their primary and defining relation to the God of the Exodus."[29] This message rings throughout Christendom, paralleling Alexander Schmemann's cosmic-redemption phrase "Christ died for the life of the world."[30] It also resembles the views of Pope Benedict XVI, of whose ecological vision a recent report reads, "an essential concept of Pope Benedict is that the ecology of nature and the ecology of man share 'the same destiny.'"[31] Indeed, for in his encyclical *Caritas in Veritate*, Benedict argues that human ecology shapes environmental ecology and vice versa.[32] For him, the human and nonhuman are bound up together in such a way that detrimental activity toward one is detrimental to the other. And even though Christianity confesses that the world is God's to save, human creatures have a part to play; this role emerges from the *imago dei*, the essential element that makes humans a creature of a different kind. Again, creation allows for difference, but this is not dualism *redivivus*; rather, creation promotes a spectrum of being. The differences between creatures exist within a greater unity of source and *telos*. Consider Bulgakov's perspective of *theosis*. He argues that all things, human and nonhuman, are sourced in divine creativity and destined for intimate communion with the divine life. He writes, "All living things instinctively strive towards grace and beauty, harmony of movement, an inner rhythm in their life."[33] For Bulgakov and others, such as Maximus the Confessor, this instinctual striving is realized by the divine-human, Christ, who demonstrates the designed role of human creatureliness and thus a portrait of an untainted *imago dei*: orientating the world toward the divine through all its works and creativity.

III

The *imago dei* means that humans are endowed with a greater level of intelligence, allowing humans to be God's viceroys within creation. Aquinas affirms this gifted intelligence because its *telos* is for humans to "see God" beatifically.[34] But intelligence is misplaced when the relationships of the world are "relationships centered upon *us*, upon our individual or group agendas, as the determining factor in how we approach persons or things."[35] For Williams, as should be clear, *the problem of and the solution to* ecological difficulties arise from human intelligence. Intelligence enables human domination, but proper, ecological intelligence, or better yet, an intelligence of creation, will enable humans to rediscover their intended relationship with the earth. Inappropriate intelligence, however, enables two harmful human moods: indifference and impatience.

Indifference and impatience make human *poesis* devastating. First, Williams notes that modern society is indifferent "to the way in which biodiversity is part of the self-balancing system of the world we inhabit."[36] Society, apathetic to nonhuman life, creates beyond its "natural" reach, caring only for human prospering. It is not uncommon to hear cultural commentators regard human happiness and progress as the chief ecological aim.[37] Indifference ignores the shared life system of humans and nonhumans, choosing to discard nonhuman life that doesn't support supposed human flourishing. Davis argues that ideologies of indifference assume that the answer to technological burden comes not "by reconsidering the course of technological 'progress,' but rather by completing it."[38] Foolishly, human indifference surmises that human societies will come up with corrective answers later via the same technological praxis. Is this not the height of foolishness? At least we can conclude that trying to possess life in such a way lacks a sense of propriety. With Wendell Berry, we can conclude that "a civilization that is destroying all its sources in nature has raised starkly the issue of propriety, whether or not it wishes to have done so."[39]

The other problem is impatience. Romanced by its ability to reduce the evidence of its reliance upon the natural world, modern society moves quickly to develop new technologies, systems, and tools. Davis quotes E. F. Schumacher, who describes modern technological fetishism as "the forward stampede" of human making without regard for the outcome.[40] Modern innovations move too quickly to account for their consequences, signifying immaturity, where maturity indicates a "prolongation of gratification" so that outcomes may be weighed more fully. This is why Bruno Latour envisions that the modern hero is the one who rushes forward in time, but with no clear end in mind. The end is an ambiguous utopia, but the main aim of his rushing forward is that of emancipation from the past. As such, the onward march of the modern "is flying backward" and thus "not seeing the destruction" that is caused by this "progress" in time. Recently, however, "a conversion, a metanoia of sorts" has occurred: the modern person "has suddenly realized how much catastrophe his development has left behind him. The ecological crisis is nothing but the sudden turning around of someone who had actually never before looked into the future, so busy was he extricating himself from a horrible past."[41] This modern technological and emancipatory impatience is well illustrated in Mary Shelley's *Frankenstein* and the contemporary film *Jurassic Park*. In the latter, Dr. Malcolm reacts to an island with genetically recreated dinosaurs with this "Frankensteinian" statement, "Yeah, but your scientists were so preoccupied with whether they *could* that they didn't stop to think if they *should*."[42] Standing in contrast to modern societal impatience are the Amish. The Amish are typically misconstrued as pure luddites, but they routinely consider the use of technologies through the patience of a communal filter. They accept or reject a tool based on communal deliberation on how it will impact their community and land.[43] This community's patience far exceeds the modern industrialized society, which for Williams transforms human *Poesis* from the perfecting work of art into toxic realities.[44]

The negligent consumerist culture of exploitation brings about this transformation. Wendell Berry, in turn, says that there are only two types of society.[45] The first, the culture of care, is represented by, but not restricted to, the Amish. It asks how humans can work together in a "place," artfully nurturing it to flourish by learning its attributes, delights, and idiosyncrasies. In contrast, the second culture of exploitation does not care about "place"; it takes what it can quickly and moves on. This attitude reflects much of American culture: manifest destiny, unreflective technology, and negligent consumerism. Ironically, the term "culture" derives from "cultivation," as the cultured man was once the gardener; now, it denotes the cultivated inner human life.[46] But even with this modern dichotomy humans still do cultivate "place"; we still intelligently create worlds to inhabit. Given this truth, Rowan Williams quotes Fulla, from Byatt's *The Biographer's Tale*: "We are an animal that needs to use its intelligence to mitigate the effects of its intelligence on the other creatures."[47]

IV

Ecological problems call for a *teleological* ecological intelligence—its aim, mutual flourishing.[48] This is why Wirzba argues that the goal of life is Sabbath rest and delight; it is a Sabbath orientation that stands contrasted to the modern constitution.[49] For the modern constitution, however, the aim of life is individualist success, which often means high levels of use and consumption, and sometimes it means exploiting certain lands and species. Sabbath, on the other hand, implies that humans value other beings and use them within a proper limit. Our present crisis calls for ecological thinking (properly, creation), a culture of care, and a renewed anthropology. First, indifference and impatience must be overcome by mature reflection on the contingency and connection that define the biosphere. Williams argues "that human reasoning in its proper and fullest sense requires an awareness of our participation in the material process of the world and thus a sense of its own involvement in what it cannot finally master."[50] This participation or union is not a mystical illusion; rather, it is the rational truth of our world that "is not a wholly detached capacity, examining the phenomena of the world from a distance, but a set of skills for finding our way around the physical world."[51] Ecological rationality is incarnational, requiring intimacy with the world, and along with Christianity calls humans to live with both feet firmly planted on the earth.[52] Simply, humans must learn how to be creatures again, for as Williams states, "not to recognize our creatureliness, our incapacity to master all the conditions under which we are becoming or shall become, is close to the heart of our unfreedom, since this refusal binds us to pervasive struggle."[53]

Next, Christian ecological thinking says that we must embrace the culture of care. By this view, understanding "comes to life when a kind of empathy and imagination is stirred by a new vision of things: intelligence alone does not generate new vision, and bare argument does not on the whole change things."[54] Only care, or better, love, brings lasting change. Indeed, orthodoxy must become orthopraxy or the soul is lost; similarly, if ecological understanding does not translate to a greater care, then a flourishing life for all is lost. This is why Williams admonishes people to love the world, reminding that "perfect love casts out fear."[55]

Much of the ecological crisis results from fear—fear of scarcity, pain, overwork, others, and even "nature" itself. What is more, this fear yields impatient and indifferent cultural artifacts. Even the majority of responses to the crisis are built on the logic of fear and guilt. Indeed, the fields of environmental justice and communications offer toxic tours and environmental justice calculators so people can see just how bad things are in proximity to where they live.[56] They are defined by doomsday predilections and guilt-causing mottos. But rarely do they invoke affection or love in their modes of communication, unless, of course, they are romanticized, picturesque idylls such as those found in works such as Thomas Kincaide paintings. The problem is that while fear motivates quickly, its impact is fleeting. Humans will change their behavior only when they actually love a person or being. And humans will only love the world by understanding what it is actually for. Williams says, "If we begin from the belief that God wants us to rejoice and delight in the created world, our basic attitude to the environment will not be anxiety or the desperate search for ways of controlling it, it will be the excited and hopeful search for understanding it and honoring its goodness and its complex, interdependent beauty."[57] Since, at least, the modern period (part and parcel to the individualism of the Reformation), Christians have focused, primarily, on God's love for mankind. Anthropological reflections on John 3:16 (it also has a possible cosmic reading) and a reduction of the gospel to "grace through faith" have been heightened, and creational readings of the Sabbath, Job, Psalms, "creation's groaning," and so on, have been diminished.

Meanwhile, for Žižek, we should simply accept our ecological burden.[58] We should sit and do nothing but reflect on our connection to an immanently defined world, because that may cause us to initiate some sort of change in behavior, albeit an ambiguous one. But for Williams, on the other hand, we must care for the world and act on its behalf, even if it cannot be saved, because it belongs to God, and God delights in it. Humans are most creaturely when they live in light of this truth and honor God's delight by delighting in life as well.

Last, for Williams, the crisis calls for a renewed anthropology of priesthood. Under the modern constitution, humans are despotic rulers separate from other beings. This Western illusion of human nature results in societal possession, dominance, and extreme usage, at least as long as our actions do not ultimately endanger future human generations.[59] Here again, we pay witness to the utter prioritization of human life over other beings; but this is not enough, for it seems, at least within the great tradition of Christianity, that God cares about the whole world. Humans are not the only beings that God created or for whom he incarnated in order to redeem. Christianity compels humans to bridge God's life and the world's life. This is why Williams writes of a priest-like relationship with the natural order: "the human agent is created with the capacity to make sense of the environment and to move it into a closer relationship with its creator, drawing out of it its capacity to become a sign of love and generosity."[60] Williams understands creation as a gift; humans possess the ability to, in freedom, align life with its true *telos*— communion with the creator. The purest form of human creativity occurs when humans are no longer seen, reductively, as *homo faber*, but as *homo adorans*. Priesthood means that humans are the only free creatures who can present the world to God on behalf of non-free creatures; this brings "non-free" members of creation into their proper orientation.[61]

Ecological rationality and care enable priestly living. For Williams, "renewing the face of the earth then is not an enterprise of imposing some private human vision on a passive nature but of living in such a way as to bring more clearly to light the interconnectedness of all things and their dependence on what we can master or understand."[62] It does not impose individual or corporate human visions, but it enacts the vision of the creator. This means learning to delight in the world as God does, even if it is something we consume,[63] as we represent the creation before the creator and vice versa. Since creation is gifted, we can affirm John Damascene's exhortation "do not abuse matter . . . the only thing that is dishonorable is something that does not have its origin from God,"[64] and uphold that salvation is for all matter, not mankind alone. Indeed, humans are responsible for acting in a way that brings flourishing to all creaturely life. Even if Žižek is right and we cannot solve ecological problems, humans still have an obligation to live gently in the world.[65]

The Art of Creaturely Life:
A Question of Human Propriety
Norman Wirzba

There appears to be a law that when creatures have reached the level of consciousness, as men have, they must become conscious of the creation; they must learn how they fit into it and what its needs are and what it requires of them, or else pay a terrible penalty: the spirit of the creation will go out of them, and they will become destructive; the very earth will depart from them and go where they cannot follow.

—*Wendell Berry, "A Native Hill," in* The Long-Legged House

Human beings have lost their creaturely nature; this has been corrupted by their being *sicut deus* [like god]. The whole created world is now covered in a veil; it is silent and lacking explanation, opaque and enigmatic.

—*Dietrich Bonhoeffer,* Creation and Fall:
A Theological Exposition of Genesis 1–3

In 1988 Jean-Luc Nancy convened a group of leading French philosophers around the question "Who comes after the Subject?" Nancy wanted to assess the status of human subjectivity after much reflection upon it by thinkers as diverse as Hegel, Marx, Nietzsche, Freud, Heidegger, Bataille, and Wittgenstein, but he also wanted to explore what such reflection looks like in the wake of a century punctuated by war, fascism, Stalinism, the camps, decolonization,

the birth of new nations, American economism, and the prolifera-
tion of (increasingly uncompelling) signs. Far from being a nihilistic
exercise in the obliteration of subjectivity or the self, Nancy wanted
to see how our thinking about subjectivity might be opened up to
fresh thoughts and new possibilities. Given numerous philosophical
critiques and a century of horror, there could be no simple "return
to the subject." We need to move forward to someone. But *who*? The
question was how to name, narrate, and receive this "someone."

In this essay I argue that, from a Christian point of view, it is
creatureliness that comes after (and before) the subject. I maintain
that creatureliness is a more faithful and compelling rendition of human life than are modern characterizations of subjectivity that have
often been uncritically absorbed by Christians. What it means to
name and narrate human life in terms of its creatureliness, however,
is anything but simple. In part this is because humanity's sinful condition, what Dietrich Bonhoeffer described as our striving to become
like god (*sicut deus*), marks an abiding rebellion against creatureliness.[1] But it is also the case that theologians have often missed op-
portunities to develop the doctrine of creation for its anthropological insights.[2] My fundamental presupposition is that creatureliness
goes to the heart of human identity and vocation, illuminating *who*
we are, *where* we are, and *what we are to do*.

My account of creatureliness will begin by developing an agrarian
picture of creaturely identity by engaging several key elements of
Genesis 2, the oldest creation story in scripture. As I develop this
account I will appropriate Bonhoeffer's suggestive theological commentary on this passage, which he presented at the University of
Berlin in the 1932–33 winter course "Creation and Sin." I will also
put this commentary in conversation with writers who have been
critical of the construct called "the modern subject," all with the
aim of opening a space for a reconsideration of the human *as* creature. I will conclude by outlining three marks of human propriety
that follow from my account of creatureliness.

Our Garden Context

It is of profound theological and anthropological significance that the earliest biblical creation story places human beings in a garden. Why this agrarian setting as opposed to some other? Bonhoeffer suggested that the setting represents a fantasy: for the Israelites, living as they did in an arid region, what could be more magnificent than a garden with rich soil, abundant water, and trees laden with beautiful and delectable fruit? The garden imagery of this story, he thought, needed to be translated into the language of today's technical world (*CF*, 81–83). We should ask, however, if Bonhoeffer's judgment is not itself a reflection of a modern, urban forgetfulness of and bias against agrarian ways of understanding human identity and life, ways that were common to most of humanity in the last ten thousand years and that were presupposed by the writers and hearers of scripture.[3] Is not the rebellion against creatureliness that Bonhoeffer powerfully describes mirrored in humanity's longstanding rebellion against the land? Perhaps the agrarian garden setting, and the practical sympathies and sensibilities it makes possible, are crucial because of their unique ability to illuminate our condition.

What does the garden scene in Genesis 2, particularly its reference to the creation of the first human being (*adam*) formed out of fertile soil (*adamah*), say about creaturely life? Most basically, it says that creatureliness is inescapably marked by need and by dependence on fellow creatures and a creator. It is easy to overlook the significance of what is being communicated here: the need and dependence that mark human life, though clearly having social and political dimensions, are first and forever experienced in bodily attachments to material bodies that are unavoidable because it is *through* them that we live at all. Genesis 2 describes human life, but also plant (2:9) and animal (2:19) life, as fundamentally and inextricably bound to and dependent upon soil. Soil is the recombinant and regenerative matrix out of which all terrestrial life comes and to which it eventually returns. As Wendell Berry, the foremost agrarian writer of our time, puts it, "The soil is the great connector of lives, the source and destination of all. It is the healer and restorer and resurrector, by which disease passes into health, age into youth, death into life. Without proper care for it we can have no community, because without proper care for it we can have no life."[4] To rebel against soil, even to neglect it, is to take a stance against creation.

Genesis 2 further shows creaturely life as ultimately dependent on God's life-giving creativity, creativity that takes an intensely intimate form as God's own breath as the breath within our own, lifting soil to an animate and *adamic* form of life.[5] Soil is never simply dirt or dirty. It is the bearer of the divine breath of life. Creaturely life, as David Kelsey says, breathes a "borrowed breath" from God. According to the Psalmist, the day God withholds this divine breath is also the day creatures die and return to lifeless dust ('*aphar*). But "When you send forth your spirit/breath (*ruach*), they are created; and you renew the face of the ground (*adamah*)" (Ps. 104:30).

The dependence described in this story is not abstract or optional. It is embodied, smelled in every breath and tasted in every swallow and bite. Appreciating it presupposes active engagement and skilled work. This is why God enlists the human creature to till and keep the garden (Gen. 2:15): because it is through the tending and serving of fellow creatures that the *adam* practically probes and potentially learns to appreciate the range, depth, and responsibilities of interdependent life.[6] According to this story, it is crucial that we keep our hands familiar with soil so that we don't forget our need and dependence, but also our responsibility to care for the bodies we live through. Human creaturely identity and vocation come together in the work of gardening. Moreover, given that God is cast as the First Gardener (2:8), we are led to think that human participation in the work of gardening is also a growing in the understanding of God's creative, attentive, patient, and nurturing ways. Gardening, in short, is the complex activity that leads us into a deeper encounter with and understanding of creation, creatureliness, and the creator's life.

My interpretation of Genesis presupposes an appreciation for how gardens are indispensable places in which insights about creatureliness can be learned.[7] Here, amidst water, soil, plant, animal, weather, and sun, gardeners work to understand as precisely as possible the character of the relationships and responsibilities that make eating and drinking, and therefore the many quotidian elements of our life together, possible. Here people discover that the sources of health and vitality are never simply "resources" awaiting our procurement, but are instead the fruit of a mysterious, fresh, enlivening power that transforms death into fertility and seed into fruit.[8] In gardens life is daily witnessed and felt to be vulnerable and fragile but also surprising and miraculous. We become good gardeners insofar as we learn to work with the powers of life that exceed our comprehension and control, even as we engage them to meet our needs. If we are attentive and honest, we begin to see that human agency sometimes destroys the health of gardens and their creatures and so needs restraint and correction.

As those who work closely with soil can testify, the power of life witnessed in gardens is a "dark" power because it so often leaves gardeners in varying states of incomprehension. Gardens are places of inexplicable fecundity and freshness, but also danger and death. Here human ingenuity and ambition are frequently revealed as floundering, contentious movements that lead to failure rather than fruit. Gardening is a form of work that perpetually undoes our knowing and unseats the gardener as the center of primary significance. It demands forms of attention, patience, and humility that, for good reason, parallel the ascetic movements of mystical quests.[9] Berry puts it this way:

Until we understand what the land is, we are at odds with everything we touch. And to come to that understanding it is necessary, even now, to leave the regions of our conquest—the cleared fields, the towns and cities, the highways—and re-enter the woods. For only there can man encounter the silence and darkness of his own absence. Only in this silence and darkness can he recover the sense of the world's longevity, of its ability to thrive without him, of his inferiority to it and his dependence on it. Perhaps then, having heard that silence and seen that darkness, he will grow humble before the place and begin to take it in—to learn *from it* what it is. As its sounds come into his hearing, and its lights and colors come into his vision, and its odors come into his nostrils, then he may come into *its* presence as he never has before, and he will arrive in his place and will want to remain. His life will grow out of the ground like the other lives of the place, and take its place among them. He will be *with* them—neither ignorant of them, nor indifferent to them, nor against them—and so at last he will grow to be native-born. That is, he must re-enter the silence and darkness, and be born again.[10]

To enter into the knowledge of their creatureliness people must live and work with the dark—that is, with an honest appreciation of their ignorance and impotence. They must learn to calm the ravenous and rapacious intellect that wants, through its knowing, to comprehend and control the world. It is through darkness that the creative light that nourishes the world can be beheld. It is in the quiet that another can finally be heard.[11]

When the Genesis 2 story is read alongside other creation stories it becomes apparent that scripture understands human life as finite, bound, and limited. Our dependence on the creator and on fellow creatures means that life is never our own or within our conceptual grasp. Who we *are*, our ontological status, is to be in need of and in relationship with a bewildering array of others, constantly receiving from them the many forms of bodily nurture and imaginative inspiration that make everyday life possible. Nothing is more indicative of this fact than our daily need to breathe, drink, and eat. To be a creature is to be incomplete, in-breathed, unself-sufficient, unable to rise and stand on one's own. Our identity is open, varied, and unfinished because it is always being worked out with the creatures we meet and the relationships we live through.

My reading of Genesis 2 shows that creatureliness means that we are always already and viscerally (through lungs and stomachs) implicated in and in-formed by others—bacteria, worms, butterflies, chickens, cows, gardeners—all of which together depend on the wild power of God as their source. Though creatures can be centers of agency in their own unique ways, nevertheless God is intimately and mysteriously present in the liveliness witnessed in their activity. Creaturely life is always life *received* from God and *inspired* and *nurtured* by others. To "be" is to be dependent and vulnerable, daily faced with the incomprehensibility of ourselves and the world in which we move. It is to be marked by potential but also always by *need*.

Refusing Creatureliness

Bonhoeffer's commentary on Genesis 2, though not focused on the agrarian dimensions I have outlined, is important because it develops an understanding of creatureliness centered on need, finitude, and limit. Referring to the placement of the tree of life *and* the tree of the knowledge of good and evil at the garden's center, he observes, "*The human being's limit is at the center of human existence*, not on the margin (*Die Grenze des Menschen ist in der Mitte seines Daseins, nicht am Rand*). . . . The boundary that is at the center is the limit of human *reality*, of human *existence as such* (*Daseins schlechthin*)" (*CF*, 86). In other words, a limit is not an obstacle or challenge that lies before us as something to be overcome and then left behind. If it were, it would be at the periphery of our lives as the domain *not yet* appropriated and internalized. Theologically understood, limit goes to the core of our being because it marks us as ones who must constantly go to the tree of life and receive life as a gift from *beyond* our own power. Ecologically understood, limit describes our condition as embodied creatures that daily draw on ecosystems and ecosystem processes for life. Limit encompasses the whole of being (*das gesamte Dasein*) and every possible disposition and manifestation of human life (*das Menschsein in jeder möglichen Haltung*). Adam recognizes and realizes himself not by overcoming the limit but by embracing and gratefully receiving it as the blessing that animates and nurtures him through life. "Adam does not know the boundary as something that can be transgressed; otherwise Adam would know about evil. Adam knows it as the given grace that belongs to his creatureliness and freedom. Adam also knows, therefore, that life is possible only because of the limit" (CF, 87).

According to Bonhoeffer's interpretation of this story it is crucial to understand that limit and need are perceived by Adam not as deficiency but as good, because he thereby acknowledges that he lives by grace rather than through the power of his own might. "The limit is grace because it is the basis of creatureliness and freedom; the boundary is the center. Grace is that which holds humankind over the abyss of nonbeing, nonliving, not-being-created" (CF, 87). The prohibition against eating from the tree of the knowledge of good and evil is not at first a temptation. As far as Adam is concerned, the prohibition pertaining to this particular tree is part of the grace of creaturely life that is already understood to be marked and enlivened by limit.

As is well known, Adam and Eve transgress the limit.[12] They appropriate and internalize it in a most graphic way: by eating. In this eating they precipitate death, for as God had said earlier to Adam, "for in the day that you eat of it [the tree of the knowledge of good and evil] you shall die" (Gen. 2:17). What does this "death" mean, particularly if we understand that creaturely life—granted as a gift rather than grasped as a possession—was marked by mortality from the beginning? Bonhoeffer is clear that the death spoken about by God is not cessation of biological existence. Instead, it is a dishonest and damaging way of existing. It is dishonest because it denies that we daily depend on others and upon God for life. It is damaging because it transforms a world of grace into an arena of competitive grasping and self-glorifying manipulation—that is, a place for the exercise of multiple idolatries.

Genuine, creaturely life is marked by the humble, grateful reception of life as a gracious gift from God and is witnessed in the responsible care of fellow creatures. Deathly "life," the existence that disrespects and violates limits, is marked by the obligation to live from *out of oneself* and is witnessed in the exploitation of others

(others matter to the extent that they can support one's living out of oneself). But this is an impossible, frustrating obligation and a fundamental self-deception, because no creature is the source of its own life. Wanting to live from and in terms set by itself, the self nonetheless recognizes—the moment it eats!—that it depends on others. Frustration leads to the rebellion that results in the death of others because their integrity and sanctity are denied in their being appropriated by Adam. In the effort to secure life and make it susceptible to his decision, Adam puts himself in opposition to the animation and nurture of God. He is unable to acknowledge and appropriately respect his life as lived *from*, *with*, and *through* others. Adam refuses grace, denies his life as a blessing, but must continue on as one defined by need. Death means "no longer being able to live before God, and yet having to live before God. It means standing before God as an outlaw, as one who is lost and damned, but not as one who no longer exists" (*CF*, 90). In seeking to secure life on his own terms, by trying to live from out of himself, Adam shows his rebellion against creatureliness and the creator. He sets in motion a history of humanity that is set *against* limit.

It would be an enormously complex task to demonstrate the many ways in which modern characterizations of the subject represent the fulfillment of this rebellious spirit.[13] What is instructive, however, is the extent to which several postmodern critiques of subjectivity are attuned to and help illuminate the formal dynamics of the Genesis story as described by Bonhoeffer. Jean-Luc Nancy, for instance, writes, "The question [who comes after the subject?] therefore bears upon the critique or deconstruction of interiority, of self-presence, of consciousness, of mastery, of the individual or collective property of an essence."[14] Nancy's reading of the modern philosophical tradition, a tradition he thinks summarized by Hegel, posits a subject that appropriates to itself, in *a priori* and (given modern technological powers) practical fashion, a world of exteriority and strangeness. Modernity moves according to a metaphysical picture of being as appropriation. Others, to use theological language, do not signify as a grace received. Instead, they appear as things waiting to be appropriated by us. In place of this appropriating subject, Nancy asks if we might envision a "someone" who is not master of itself and others but instead "comes indefinitely to itself, never stops coming, arriving," thereby suggesting an identity marked by openness to a genuine other, a genuine limit (*WCAS*, 7).

In similar fashion, Michel Henry reflects upon the technological character of modern culture. "Technology consists in the unconditional subjugation of the Whole of being, which becomes the Object, to man, who becomes the Subject—the Ob-ject of the Subject, then, dis-posed before him and disposed of by him, at his disposal therefore, having no other end than this being at the disposal of, subject to tallage and corvée as the serf of this new Lord" (*WCAS*, 158). For Henry it is Kant who best summarizes the modern subject as the transcendental self who appropriates all beings to itself through various acts of representation: "For the subject is nothing other than this: that which in making appearances appear, in this same gesture, makes be everything that is" (*WCAS*, 163). To represent others is to bring them within a horizon of meaning and significance that is determined at the outset. Henry insists that there is dishonesty and deception involved here because this subject, though perhaps responsible for the appearing/signifying of all others, cannot bring about its own existence. The subject that actually exists in its world does not exist as the result of its own representation of itself. What is needed, thinks Henry, is a way of characterizing the post-subject as someone appearing in an ek-static movement—a coming from beyond that is also a witnessing to the transcendence of others—that takes him or her beyond the security and sameness of representational consciousness.

This postmodern desire to open the self to the one who is genuinely other was further reflected in Jean-Luc Marion's response to Nancy's question. In "*L'Interloqué*" Marion offered an analysis of an encounter with a genuinely other person. In this encounter the subject's self-mastery is destabilized and decentered by a claim that is made. As spoken to or interpellated by another, the first appropriate response is, "This is me!" It is a response made without the mastery or confidence normally assumed by the modern subject. "I experience myself being claimed, that is, called upon in the accusative—interpellated as suspect and not as subject, named in the accusative and therefore dispossessed of any nominative function. The interpellated *me* marks the absence of any constituting *I*, under the—in this respect, totalitarian—authority of the claim" (*WCAS*, 243). For Marion, the claim of the other upon me spells the "disaster" of the I.[15]

Nancy, Henry, and Marion each worry about the self's desire to secure itself and the world in terms established by itself. As autarchic and autonomous, the self is not genuinely open to or receptive of a genuine other. It acknowledges no limit at the center of its life. It can only perceive limit (temporarily) at the margins and as a reality to be overcome.

If we return to the Genesis story we can see that the rejection of limit is ultimately a rejection of God. Consider Eve's encounter with the serpent, who asks, "Did God say, 'You shall not eat from any tree in the garden?'" (Genesis 3:1). Bonhoeffer has no interest in assigning blame to the woman. Nor does he think it fruitful to ask where the serpent comes from (the Bible, he says, does not aim to explain something like the origin of evil). What he focuses on is the new, decisive possibility that the serpent's question is raised within the human being: "through this question the idea is suggested to the human being of going behind the word of God and now providing it with a human basis—a human understanding of the essential nature of God" (*CF*, 106). For Adam and Eve the serpent's question becomes a "godless question" not because it is a question *per se*, as if questions were impermissible, but because Adam and Eve now place themselves in the position where they are the judges over how questions are to be answered. There are questions that are in the service of loving and learning. But there are also questions that aim to establish the self as the authority by which others are to be judged and understood. Bonhoeffer thinks that the encounter with the serpent brings about this latter kind of questioning and response. Adam and Eve do not respond by saying "Here am I," thereby opening themselves to the other. Instead they ask "Did God say?" thereby establishing themselves in a position of power and mastery. Rather than submitting to God's word, they exalt themselves to a position *sicut deus* (like god), and now live in open rebellion against God. "Humankind is now sicut deus. It now lives out of its own resources, creates its own life, is its own creator; it no longer needs the Creator. . . . Adam is no longer a creature. Adam has torn himself away from his creatureliness" (*CF*, 115).

For Bonhoeffer the authentic human creature is the one who accepts life as an encounter with others and as a grace that comes from beyond the power of human knowing. For Marion the *interloqué* is the one addressed (*der Angesprochene*) and claimed by another. Several features of the *interloqué* bear noting: this self is not autonomous because it is always already compelled to be in relation;[16] this self lives through surprise rather than through itself; and, this self is always subject to the judgment of the other. "The *interloqué* provides the beginning—the most basic, hence the first, determination—that abolishes the subject: selfhood is initially wounded by the very fact that, before the self can constitute itself, the claim has already exiled it outside its 'mineness.' The wound that originally tears selfhood obscurely manifests the origin itself—the *interloqué*. Before ever knowing by what or by whom, the *I* surprises itself, as *interloqué*, and has always done so" (*WCAS*, 244–45).

Nancy, Henry, and Marion present critiques that, in various ways, characterize modern subjectivity as an imperial, totalizing presence in the world. In the work of annexing and appropriating the world—what Bonhoeffer described as humanity's sinful rebellion against creatureliness—others are repeatedly transgressed and violated. Though continuously active, this is a self that, according to Emmanuel Levinas, is nonetheless asleep (might we not also say, following Bonhoeffer, "dead"?) because not alive and responsive to another in its singularity and transcendence.[17] For Levinas "the very life of the human" is in the unsettling of the mastery of the self in the approach of a genuinely transcendent other who calls and inspires the self to a life of responsibility.

Embracing Creatureliness

The Genesis story we have been following indicates that human creatureliness is worked out in care and companionship. Besides being invited to take care of fellow creatures by "tilling and keeping" the garden, God says, "It is not good that the man should be alone; I will make him a helper as his partner" (Gen. 2:18). Bonhoeffer observes that elsewhere in the Bible God alone is designated as a partner and a help to human beings. We should, therefore, be astounded that animals, perhaps because sharing the same soil-based body and divinely breathed vitality, are presented to Adam as potentially fulfilling this position. "At the point where God wishes to create for the human being, in the form of another creature, the help that God is as God—this is where the animals are first created and named and set in their place" (CF, 97). Adam names the animals, thereby establishing a relationship with them, but none of these relationships attain a level of a genuine partner and helper.[18] Why this is so we are not told.

God then causes a deep sleep to come over Adam, during which time God removes a rib and creates another human creature from it. This creature is presented to Adam, who calls her woman (ishshah) because she comes physically from him (ish) and is "bone of my bones and flesh of my flesh" (Gen. 2:23). According to Bonhoeffer it is significant that this woman is created while Adam is asleep because this reinforces that she exceeds his expectations and preparation. Though fashioned from his flesh, the woman is decidedly neither an extension of him nor the result of his decision. She is a limit, a creature with her own integrity that emerges out of the darkness of sleep. "That Eve is derived from Adam is a cause not for pride, but for particular gratitude, with Adam. Adam does not infer from it any claim for himself; instead Adam knows that he is bound in a wholly new way to this Eve who is derived from him. This bond is best described in the expression: he now belongs to her, because she belongs to him" (CF, 97). Though the woman is clearly a limit to Adam, she and he exist in a needful and necessary relation to each other, a relationship described as mutual belonging and the sharing of one flesh. The two do not merge or blend into each other so as to abolish their individual creaturely identities. The belonging that characterizes their life together is based precisely on their being different from each other.

Mutual belonging and the companionship it makes possible reveal a profoundly important way of living with limit. This way of living Bonhoeffer calls love. Prior to the creation of the woman, Adam related to limits with the understanding that they were to be received as God's gracious gifts. Adam received the gift of the other with faith but not yet with love. "The Creator knows that this free life as a creature can be borne within its limit only if it is loved. . . . The helper who is a partner had to be at once the embodiment of Adam's limit and the object of Adam's love. Indeed love for the woman was now to be the human being's very life (in the deepest sense of the word)" (CF, 98). The woman is for Adam a unique other or limit because she is made from his body and so is intimately related to him. This intimacy, however, entails a reciprocal love by her for him because she knows herself to be drawn from him, indeed carries his body within her. The reciprocal love of the man and woman is foundational because it makes possible a life that can bear limits. It is love that will keep the human creature from transgressing, violating, and appropriating another because it is love that enables the lover to make room for the beloved to be itself. Love creates the space and the freedom for another to be. Without love there is the danger that the other will be perceived as a threat or as something to be hated. When this happens, shame enters the world.[19]

Erazim Kohák has shown that the belonging that characterizes the relationship between Adam and Eve is not confined to the human realm. Living on the land, patiently and with affectionate regard for it, gradually produces the sense that just as the land belongs to us we also belong to it. (It is not insignificant that agrarian traditions describe the bond between human beings and the land and its creatures by using the marital language of "husbandry.") Eating food grown on one's place, heating oneself by its energy, allowing oneself to be inspired by its potential and beauty—all occasions that join our flesh to the flesh of the world so that it can rightly be said that we become "one flesh" with it—reveal a fundamental deception in all claims to possess land outright. Kohák argues that the concept of possessing operates at a formal level that often denies the life-giving bonds that exist between us. "The bond of belonging that grows up over years of life, love, and labor is the most basic truth of being human in a world."[20] Labor, rather than contributing to an accumulation of land understood as private capital (as John Locke thought), leads to an appreciation of the sanctity and grace of the world insofar as this labor is inspired and directed by love. To say that another belongs to me is not to make a possessive claim. It is, rather, to indicate that without him, her, or it my life would be diminished. Recognizing the other's value and integrity is therefore also an invitation to commit to this other's well-being.

After describing the man and the woman's life as the love of mutual belonging and becoming one flesh, scripture adds, "And the man and his wife were both naked, and were not ashamed" (Gen. 2:25). Bonhoeffer interprets their nakedness to mean their innocence before and their obedience to each other. In their primordial state the man and woman do not face each other with fear, remorse, or as a threat because their life together is one in which the integrity of each other is affirmed and served. "Where one person accepts the other as the helper who is a partner given by God, where one is content with understanding-oneself-as-derived-from and destined-for-the-other, in belonging-to-the-other, there human beings are not ashamed" (CF, 101). The ultimate meaning of nakedness, we can say, is to be found in the act of self-offering to another, an offering in which nothing is hidden from the other and nothing is kept for oneself except insofar as it might be shared. Once shame appears, once the pornographic desire to objectify and control the other takes hold, it can only be overcome by the forgiveness that restores unity and communion with others. Forgiveness acts as a kind of "unclothing," removing suspicion, hatred, envy, and alienation, and so restores people to a reconciled condition in which they can stand naked before others without shame.[21]

Bonhoeffer's interpretation of this story shows that creatureliness is something not to be endured at all, or perhaps only temporarily (and while awaiting escape to some other-worldly heaven).[22] We can say this because creaturely life at its most profound realization leads to the loving embrace of the other, an embrace that does not stifle or diminish others but instead nurtures them to more fully become themselves. In the belonging to and service of others love is revealed as the hospitable gesture that takes from what one has received (even one's own body) and offers it to another. Human creaturely life, life that is without shame, presupposes that I can relate to others in such a way that my life—what I need, desire, and enjoy in life—makes no sense apart from the belonging and fellowship of life *together*.

Having briefly outlined the contours of human creaturely identity, we can now turn to some of the practical implications that follow from this account. Responses to Nancy's question "Who comes after the Subject?" show that we cannot think about human identity without also thinking about human propriety. To ask about *who* we are is also to ask about *how* we are to live *where* we are. The question "how?" compels us to think about propriety, about how we "fit" and comport ourselves within the world. The question of "who?", in other words, is not theorized in the abstract. It is worked out, discovered, and revealed in patterns of life practiced in the world.

As Nancy and colleagues examined the records of modernity they saw humanity's inability to fit harmoniously within a world of others. Violence, exploitation, subjugation, appropriation, neighborhood neglect, conspicuous consumption—these were the patterns of life made evident in histories of war, colonialism, sexism, fascism, and ecocide. Had they been attuned to the analyses of Bonhoeffer, they might have concluded with him that we have yet to appreciate and implement the sort of relationships that respect, serve, and cherish the mystery and the grace that others are. The desire to be autonomous and autarchic, the decision to live from out of oneself, appropriating the world at will—all movements characterized by Bonhoeffer as the desire to be *sicut deus*—have led to an unrelenting violation of others and the steady, systemic degradation of the world.

We should pause for a moment so I can be clear about what I mean by "world." Reading Nancy, Henry, Marion, and Bonhoeffer, it is clear that they are preoccupied with the social world, a highly urbanized world at that. When they speak of a limit to the self and an opening to the other, they are almost always referring to a personal other. What has been overlooked by them, and by vast stretches of our philosophical and theological histories, is the sense of limit that is fundamental to a breathing, eating, and drinking body. Whereas social limit refers us to political structures and *de jure* forms of dependence, embodied ecological limits point us to *de facto* forms of dependence that are fundamental, necessary, and inescapable. Of course, political structures quickly shape the forms ecological relationships take (consider the various ways property and land management have been configured across time). But in overlooking the material world of creation we will end up with an impoverished understanding of human creatureliness and creaturely responsibilities. It is no accident, I would argue, that the myopic focus by philosophers and theologians on strictly inter- and intra-human affairs has led to degraded fields, forests, waters, and sky. To conclude this essay I will therefore briefly develop what I take to be three essential marks of creaturely propriety, marks that join us to other people *and* to the soil on which we all depend: the acknowledgment and embrace of the goodness of limits; the reception of life as a gracious gift; and hospitable service to others.

First, creaturely propriety begins with an acknowledgment of the goodness of limits and the refusal to attempt a life *sicut deus*, a life unfettered and unencumbered by limits of any kind. The embrace of limits, rather than simply their toleration, is crucial because it makes possible an honest estimation of ourselves as embodied, communal creatures belonging to, living within, and nurtured by a vast membership of creatures ranging from bacteria to bees to bee-keepers. Individual life is a contradiction in terms. Individualism is a violation of propriety. To acknowledge others as limits is to realize that we must exercise restraint in our relationships lest we become rebels upon the earth. To appreciate limits as good is to affirm that life is a miracle having its source and vitality in powers that exceed our expectations and control. Insofar as we sense the miraculous character of life, we learn to face the twin temptations of hubris and despair: hubris, insofar as we think we are the source and center of life, and despair, insofar as we think we are meaningless accidents appearing only momentarily in a random universe.[23]

To describe life as a miracle or as a mysterious grace is to call into question the modern project that would comprehend the world as a mechanism and a resource to be manipulated at will.[24] Besides leading to a feeling of human alienation from the world (Kohák describes how a scientific picture of an essentially dead and valueless world requires us to bracket our embodied and life-giving relationship to earth, which in turn leads to a philosophical picture of humans as strangers "thrown" into a world),[25] a characterization of the world as a machine trains us to presume that it is also explainable, predictable, and within our control. What this characterization fails to do is note the fundamental incomprehension that circulates throughout our living, an incomprehension Henry Bugbee described as the inescapable wildness at the center of our being: "The more we experience things in depth, the more we participate in a mystery intelligible to us only as such; and the more we understand our world to be an unknown world. Our true home is wilderness, even the world of everyday."[26] To know this home, however, presupposes that we have become active, patient, and humble participants within it. For humanity to live appropriately and well, *adam* must remain close to *adamah*, constantly keeping the soil in mind and in his hands.

It was not inevitable that we should reduce life and creation to the predictability of a machine, for, as Thomas Carlson has shown, thinkers at the origins of modernity were also known to be captivated by the world's and the creature's incomprehensibility and by the human being's inability to take secure possession of itself. Pico della Mirandola, for instance, argued that human agency and creativity in the world revealed our fundamental indeterminateness insofar as we are understood to be makers made in the image of an incomprehensible God.[27] Pico reflected a long mystical tradition that presupposed a world that exceeds conception because it is made by a God who exceeds comprehensive grasp. Though God is revealed in what is made, the ground of creativity itself remains incomprehensible. The consequence of that realization is that we do not know who we are or what the world finally is. For Giordano Bruno, an infinite God entailed an infinite creation. "As creatively active within the infinitely dynamic tissue that constitutes and exceeds the self, this relational self lives and moves always by means of a vision or knowledge that implies blindness and unknowing. Its knowing and doing will never exhaust the tissue of the real, but as productively participant in that tissue, such knowing and doing imply an infinite responsibility" (TII, 117). When we acknowledge the miraculous, wild, infinite, and incomprehensible character of life, we also open a space in which amazement, enchantment, and renewed questioning can take hold.

Second, creaturely propriety entails the reception of life as a gracious gift. A great deal has been written about the meaning of a gift and whether or not a gift can ever be given or received. Robyn Horner, in her philosophical and theological study of the gift, shows that treatments of the gift, especially those influenced by the work of Jacques Derrida, refer to the aporetic character of gift-giving. For instance, if a gift is part of an exchange called into being by obligation, guilt, a desire to influence, or payment for services, then it is no longer a gift. Moreover, the moment a gift is identified, it seems to have become something else—a commodity, a prized possession, a trophy, or a reminder of a debt. This is why Derrida says, "For there to be a gift, there must be no reciprocity, return, exchange, countergift, or debt."[28] Horner is right to conclude that the Derridean aporia of the gift does not mean that giving has therefore come to an end. Rather, it is to come to the recognition that we can never comprehend a gift as such, nor can we ever fully understand what we are doing when we offer a gift.[29]

As we have described it, to be an embodied creature is necessarily to find oneself placed within, nurtured by, and responsible to a world of others that come from beyond our planning or control. It is to appreciate that life is not a possession but a membership of receiving, sharing, and offering again. That we eat, drink, and breathe means that we must constantly receive, dimly perceiving that every bite, gulp, and breath implicate us in life-and-death dramas that exceed our best efforts to understand.[30] Though we may call food a "gift," this in no way entails that we have comprehended or exhausted the significance of what we so name. That we name creation, even life itself, a gift does not mean that we are in a position to fully understand what we mean.[31] What we are communicating with gift-language, however, is that we do not live alone or from out of ourselves. We are creatures bound to soil and fellow creatures, altogether in need of the grace of life. The language of "exchange," though useful in a business world of self-possessed consumers negotiating commodities, is not appropriate in a creaturely world because this world, the world defined by need, embodiment, limit, eating, death, pain, beauty, and warmth, moves not as a series of business transactions but as a sphere of mutual belonging and responsiveness.

The creation story in Genesis 2 showed us that Adam (it is appropriate to capitalize "Adam" now because this is post-gender differentiation) needed to learn to live with limits. He needed to understand that limits are good rather than a threat and that the most authentic realization of creatureliness is demonstrated in his love for another. This love we described as Adam's nakedness before Eve, his offering of himself to her. It is appropriate to call this self-offering a form of giving, but we need to see that the context of his offering is born out of vulnerability and the incomprehension of being inspired, informed, fed, and met by countless others. He does not know who he is giving himself to, nor does he have possession of himself. What he glimpses is that he is marked by need—need for breath, need for food, need for companionship, need for help: a fundamental need for life—and that the blessings and the pains, the responsibilities and the meanings of his needs can only come to light as he gives himself to the becoming of one flesh with Eve and one flesh with his nurturing place. Adam's most fundamental and abiding creaturely task is to be a witness to the wide scope of his need. It is to demonstrate with the offering of himself that he lives only because he always already receives.[32] It is to testify to a world marked by membership and belonging, but also mercy and forgiveness.

It is important to stress mercy and forgiveness because it is in terms of these that we more fully understand a third mark of creaturely life: the hospitable welcome of and service to others. Hospitality to others is rooted in mercy because, without forgiveness, work done for others might have the effect of binding them to us. Forgiveness, as the parable of the master/servant in Matthew 18:21–35 shows, releases people from crippling debt and liberates them to experience the fullness of life. Hospitality informed by such forgiveness makes room for another to grow and become itself. To appreciate what this means, we can return to the garden, for it is in gardening work that the practical dimensions and responsibilities of hospitality come into view.

Recall that Adam's gardening work is not simply for providing food. By immersing his hands in soil, by committing himself to the growth and flowering of others, Adam is learning both who he is as a member of creation and how he can best live where he is—that is, in ways that are a blessing rather than a curse to fellow creatures. Remembering that God is the primordial gardener who creates the world through gardening, Adam, by learning the skills of gardening, is learning to participate in God's life-giving, life-sustaining, life-celebrating ways with the world. God creates by "making room" and by creating the conditions in which others can freely become themselves. The whole of creation can thus be described as a performance in hospitality.[33]

Though gardening is clearly marked by active engagement—soil is prepared, seeds are planted, plants are watered and protected, plots are weeded—gardening is also marked by the gardener's withdrawal and restraint. A gardener cannot simply impose her will upon the garden. She must be attentive, patient, humble, and so learn to attune her desire, her expectations, and her work to the needs of the garden. A gardener, in other words, gives herself to the garden so that the garden can flourish. The pattern for this self-giving, says Rowan Williams, is none other than the triune, creating God: "The God who creates a world of freedom, a world that is itself, is a kenotic God, a self-giving, a self-emptying God whose being is for the other." Insofar as creatures are wise, they participate in this divine life of self-offering: "to live in wisdom is to live in and by this energy of dispossession and outpouring."[34] Just as God the gardener withdraws to make room for the world, all the while nurturing it, so too hospitable creatures withdraw to make room for the other as welcome guest, all the while offering nurture and help.

Love as the hospitality that makes room for another is extraordinarily difficult. To move into it requires an imagination and a set of skills that multiple traditions of thought and work, tuned as they are to control, prediction, possession, and comfort, actively resist. Recalling the naked vulnerability out of which Adam offered himself to Eve, who today wants to offer him- or herself to a world punctuated by violence and abuse? Recalling the naked Jesus hanging on the cross—Jesus the new Adam, the one who shows us human creatureliness in its fullest and most abundant form—who wants to empty him- or herself to the point of death? The gospel witness puts the matter succinctly: the hope of life rests on life's perpetual being given away. "Very truly, I tell you, unless a grain of wheat falls into the ground and dies, it remains just a single grain; but if it dies, it bears much fruit" (John 12:24).

A life of self-offering opens a new relationship to the world, a relationship that seeks to engage creation on *its* terms rather than our own. This is why those who embark on a path of creaturely wisdom must learn the particularizing skill that refuses to take another for an abstraction and resists the exploitive desire that reduces others to a moment within a self-serving plan. Berry puts it well when he writes:

the human necessity is not just to know, but also to cherish and protect the things that are known, and to know the things that can be known only by cherishing. If we are to protect the world's multitudes of places and creatures, then we must know them, not just conceptually but imaginatively as well. They must be pictured in the mind and in memory; they must be known with affection, "by heart," so that in seeing or remembering them the heart may be said to "sing," to make a music peculiar to its recognition of each particular place or creature that it knows well. . . . To know imaginatively is to know intimately, particularly, precisely, gratefully, reverently, and with affection.[35]

A gardener's imagination represents one of the deepest kinds of immersion into the world. In this movement we put ourselves in touch with the pain and suffering of the world.[36] In the work of hospitable gardening we are enabled to see where the work of care and justice are yet to be done. Insofar as we learn the disciplines of creaturely propriety we join, hopefully harmoniously, with the vast membership of fellow creatures, sharing in the divine hospitable work that nurtures, protects, and liberates creatures into Sabbath rest and delight.

Face of Nature, Gift of Creation: Thoughts Toward a Phenomenology of *Ktisis*

Bruce Foltz

I

Words matter. This is known to philosophers. And to poets. It is known to theologians. And to jurists, as well. Words matter in different ways. And to different degrees.

A certain word, for example, might exercise such power in a poem that it serves as a key word. Thus, in the earliest of his Hölderlin essays, Heidegger suggests that the word *nature* is a key word (*Leitwort*) in Hölderlin's poem "As When on a Feast Day. . . ." Yet Heidegger immediately adds that this key word *nature* lags behind, or falls short, of what the poet more aptly calls "the 'wonderfully all-present'"—falls short because nature designates merely "a particular realm of being": nature as opposed to art, nature in contrast to spirit, or nature as distinct from what is supernatural. Hölderlin can employ this inadequate and incongruous word, Heidegger continues, only because it resonates with the concealed truth of something other, of another word that he calls an "originary fundamental word"—that is, the word *physis*, understood in what Heidgger believes was its ancient Greek meaning.[1] Thus, beyond the *Leitwort* of a poem, we can speak of a word more important in its resonance and power, a fundamental word or *Grundwort* so originary or primal that *even when it remains unspoken*, it can allow something hidden to be experienced, something ancient to become present—it can put into play something for which we have not yet found the right word.

Physis is undoubtedly a *Leitwort* or key word for Heidegger's own writing. But beyond this, the sense of nature (and of being) that it evokes has increasingly captured the imagination of philosophers and theologians in search of more adequate ways to think and talk about the natural environment. Devall and Sessions, in their seminal book *Deep Ecology*, devote an entire section to Heidegger as one of the principal sources for deep ecology.[2] Twenty years later, in the collection on Continental Philosophy and the Environment that I co-edited with Robert Frodeman, among some twenty articles there are far more references to Heidegger than to any other philosopher—over twice as many as for Merleau-Ponty, his nearest competitor. Nor should this be surprising, given the dramatic and invigorating perspectives opened up by rethinking nature as *physis*—as not just an ontic aggregate of entities, but the ever-arising, all-embracing, captivating power of self-emergence that at the same time retains its own reserve in the darkness of earth.

By comparison, the word *environment* is philosophically shallow and inadequate, falling far short of what needs to be said. And its uncritical use immediately embeds us in a tangle of scientific concepts and presuppositions. As Heidegger had noted already in the 1920s, environment *(Umwelt)* in its biological sense overlooks the *relational* character of interworldliness and draws instead upon what he calls "the fundamentally misconceived idea that the animal is present at hand, and then subsequently adapts itself to a world that is [itself] present at hand."[3] And the more recent cybernetic concept of environment as self-regulating system, composed of information bits within feedback loops, is even more problematic and misleading. In contrast, the ancient Greek concept of *physis*, as retrieved and elucidated by Heidegger, is charming and compelling, restoring to us the enchantment of nature that Marx and Weber had told us was lost forever; it promises to repatriate us to the "wonderfully all-present" self-emergence obscured by the urban labyrinths within which we have become lost, haunted, as Camus had maintained, by the Minotaur of history, and bereft of the healing, redemptive power of the sun and sky and breezes.

We mourn this lost magic, long for the awakening of nature that Hölderlin celebrates—for the moment at the break of day, after the storm, "when out of the hot night the cooling lightning flashes had fallen . . . [when] the fresh ground becomes green and with the gladdening rain from heaven the grapevine drips, and gleaming in quiet sunlight stand the trees of the grove."[4] Here, in Heidegger's words, "*physis* is an emerging and an arising, a self-opening, which while rising, at the same time turns back into what has emerged, and so shrouds within itself that which on each occasion gives presence to what is present."[5]

"But now day breaks!" sings the poet, "I awaited and saw it come, and what I saw, may the holy be my word."[6] "In awakening," Heidegger adds, "nature unveils her own essence as the holy." The call for a re-enchantment of nature, for a return of our experience of it as *physis*, must therefore be a simultaneous invocation of a return of the divine, in the wake of whose departure we remain in a state of mourning. The retrieval of something more original, more *originary* than either the modern sense of nature or the Scholastic sense of *natura*, would thus be bound up with a retrieval of the holy, perhaps yet another way of naming for which we hunger and yearn.

The current depth and pervasiveness of this longing, about which Heidegger reflected more than a half century ago, is notably evident in a recent, popularly written book by two professional philosophers. *All Things Shining*, by Hubert Dreyfus and Sean Kelly, argues that a re-enchantment of the world is imperative if we are to lead what they characterize as "intense and meaningful lives."[7] "The world doesn't matter to us the way it used to," they begin, echoing Nietzsche. "The world used to be, in its various forms, a world of sacred, shining things. The shining things now seem far away." How, they ask, can we then manage "to lure back the shining things, to uncover the wonder we were once capable of experiencing and to reveal a world that sometimes calls forth such a mood?"[8]

Exploring several texts, the authors devote special attention to Homer and his deity-saturated world and to Herman Melville's fascination with paganism and polytheism, ending with a remarkable conclusion that echoes Nietzsche's attempt at overcoming nihilism. In a final chapter called "Lives Worth Living in a Secular Age," they advance a new polytheism that would be attuned to the sacred meanings that still radiate from many quarters in our daily life, if only we cultivate the skills to discern them and the sense of gratitude to honor them—from the natural world, of course, but also from the excellence displayed in professional sports or even the careful ritual of preparing and drinking a fine cup of coffee in the morning. Nietzsche—who had scolded his age by admonishing, "two thousand years and not a single new god!"—tried to revive the worship of a single deity: Dionysus, god of intoxication and ecstatic affirmation of life. Dreyfus and Kelly argue for a potentially *unlimited* pantheon, subject only to our ability to appreciate and indeed worship its denizens, not excluding the god of cappuccino. "We can become receptive to a modern pantheon of gods," the authors proclaim, "and we can lure back the gods of old." Their conclusion? "This contemporary Polytheistic world will be a wonderful world of shining things."[9] *Physis*, wonderfully all-pervasive, can be discerned even in the gently frothing *crema* on a fine cup of espresso. As Heraclitus remarked long ago, there are gods here as well!

Surely the ancient Greeks, like all early peoples, inhabited a god-infused, and in some ways a god-infested, world! But was this pervasive sense of "the holy" really named by the word *physis*? And when Heraclitus said that *physis* inclined toward being hidden from view, was he really speaking about some deity hunkering amidst the embers of his little stove? Or was he alluding to the *Logos* itself—also identified by Heidegger as one of the great *Grundworten* of the West—of which Heraclitus notes that it is hidden and easy to overlook, despite its ubiquity. This hidden *Logos*—whose uncover-

ing demands of us a certain purification or *askesis*, a "drying" of
the soul—is surely for him something higher than any ember deity,
to which he perhaps alludes in the stove anecdote with a certain
irony—a possibility suggested by his irreverent allusions to Dionysus? Does *physis* here characterize in any important way the *dai-
monic*, numinous character of the gods as the Greeks encountered
them, a sense of divinity that the pre-Socratics had already begun to
question? And if *physis* here does not name the holy, does it designate self-withholding self-emergence, either?

Or we may look to the ancient Greek poets, where this sense of
physis should be most readily discernible. But here, too, the sense
of the natural world that we encounter hardly resembles the dynamic sublimity that Heidegger characterizes as *physis*. Schiller had
already noted this in his *Naïve and Sentimental Poetry*, observing
that in their unsentimental descriptions of "the scenes and characters of nature," the Greeks "showed no more involvement of [the]
heart than in [describing] a tunic, a shield, . . . furniture or any mechanical product." Indeed, Schiller argues that "in his love of an
object, [the ancient Greek] does not seem to make any distinction
between those which appear of themselves, and those which arise as
a result of art or the human will."[10] When Homer, for example, ex-
tols in the *Odyssey* a certain island, it is because it is "a good land,
rich in herds, rich in flocks, full of wine, abounding in wheat." And,
of course, Hesiod's *Works and Days* is just as practically oriented
in its approach to the natural environment and the necessities it
imposes upon us. The exceptions are just as telling. When Pindar
describes lands and landscapes, he either evokes their mythic past
or celebrates the calm eternity, the dependable regularity exhibited
by the natural world. This pagan sense of a serene and ever-renewing
continuity that endures outside the city walls is invoked more recently by Camus in his unjustly neglected lyrical essays.

Seeking refuge from the insanities of war and history and politics,
Camus reflects on the Mediterranean, and specifically Greek, affinity for the measured, calmly indifferent loveliness of the sun and

sea, praising it as the only eternity upon which we can depend, our only lasting source of solace. Yet "we turn our back on nature, we are ashamed of beauty," he laments in "Helen's Exile." "The world has been deliberately cut off from what gives it permanence: nature, the sea, hills, evening meditations." Citing Heraclitus's warnings against excess, Camus adds, "nature is still there, nevertheless. Her calm skies and her reason oppose the folly of man."[11] It is this measured, moderate, "classical" sense of the beauty of nature—or else pragmatically oriented discourses—that we typically encounter in ancient Greek literature, not the vibrant, dynamic, eventful sense of *physis* that Heidegger believes to have been at play.

As for the word *physis*, the most exhaustive study in English of its ancient Greek usage joins earlier scholars in maintaining that Heidegger's exegesis is unsound: that *physis*, from its earliest usage, does not mean birth or inception, emergence or arising, but "growth," and moreover growth understood teleologically and normatively—"the (completed) realization of a becoming," "the nature of a thing as it is realized, with all its properties."[12] In other words, more or less what the English word *nature* means in its ordinary sense of "the nature of something or other." Or better yet, what Kohák—wishing to designate "the *integrity* of the nature that surrounds the wall-less space of a tarpaulin on a rainy summer day"— calls "the incoercible, intangible something that makes a being the kind of being it is."[13]

Are Heidegger's long reflections on *physis*, then, mere fabrications, wishful thinking—perhaps a belated resonance of the enthusiasm with which youthful Tübingen seminarians had once embraced ancient Greece, having found little enchantment in either the Roman civilization or the Christian theology they had inherited? Or is there instead some other source, even some other *Grundwort*, upon which Heidegger depends, perhaps without understanding its origin—perhaps even one that remains unspoken? I want to suggest that there is another fundamental word animating Heidegger's exegesis—indeed, another ancient Greek word, but one that Heidegger consistently overlooks in favor of its Latin translation. The Greek word is *ktisis*, and the Latin word that translates, and I believe comes to obscure it, is *creatio*.

II

In *Olympians* VII, Pindar recalls an ancient tale in his depiction of a seaside landscape: "that deity whose hair is gold spoke of sailing: from Cape Lerna's shore straight to these meadows in the sea: here, once, the god of gods had soaked a city in golden snowflakes."[14] It is a lovely, indeed an *enchanting* passage, but it conveys not a mode of seeing, but a kind of remembering—not the vision of a living landscape, but the recollection of a *topos* in mythic space, a storybook place animated not by any living experience of nature, but by the charming memory of what had happened, long ago, once upon a time. It evinces not *physis*, but *anamnesis*.

Let us contrast this with another poetic vision of the sea. It is from Psalm 103 in the Septuagint, which is chanted at every Orthodox Vespers. Here, at the setting of the sun, the faithful evoke that great cycle within which the living God, the eternal Logos—who covers himself "with light as with a garment" and "rides upon the wings of the wind"—shows not just that he is lord of the waning day, but that even the dark abyss "is his mantle"—that is, that he creates the night as well, the approaching time when "all the beasts of the forest go forth . . . seeking their food from God"; that he makes the moon to mark the seasons; and that the sun, which knows the time for its setting, will soon rise again, marking the time for the forest animals to retreat to their dens, while man goes forth once again unto his labor until the evening and another vespertide. And it is at this point that the Psalmist exclaims:

> How magnified are thy works, O Lord! In wisdom hast Thou made them all; the earth is filled with thy creation [*ktisis*].
>
> So is this great and spacious sea, therein are things creeping innumerable, small creatures among the great.

There go the ships; there this Leviathan, whom Thou hast made to play therein.

All things wait on Thee, to give them their food in due season; when Thou givest it them, they will gather it.

When Thou openest Thy hand, all things shall be filled with goodness; when Thou turnest away Thy face [*prosōpon*], they shall be troubled.[15]

The somewhat breathless commentary on this passage by St. Cyril of Jerusalem is indicative of power of this text:

Who can describe the beauty of the fishes therein? Who can describe the greatness of the whales and the nature of the amphibious animals, how they live on the dry land and in the waters? Who can describe the depth and the breadth of the sea or the shock of its tumultuous waves? [Yet] the sea stays within its confines because of him who said: "This far shall you come but no farther, and here shall your proud waves be stilled."[16]

Appropriately, St. Cyril concludes by citing from that remarkable passage in the Book of Job, where God narrates his creation of the "foundations of the earth": "I shut up the sea with doors when it burst forth. . . . I said, 'This far you may come, but no farther. And here your waves must stop'" (Job, 38: 8, 11).[17]

Heidegger states, "*Physis* as emergence and arising [*Aufgehen*] can be experienced everywhere, e.g. in celestial phenomena (the rising of the sun), in the surging of the sea, in the growth of plants, in the coming forth of man and animal from the womb."[18] And it is here, I want to argue, in these dramatic and dynamic lines of psalmody, and in innumerable others throughout the Wisdom Books of the Bible, and not in the measured, placid, timeless—and often ethereal and otherworldly—verses of ancient Greek poetry, that the roots of this sense for nature that Heidegger wants to retrieve can be found. *Ktisis* is just this ever fragile, perpetually enacted emergence and renewal and sustenance of heaven and earth out of the unthinkable nothingness *from* which they are ever brought forth and *to* which they always incline, a creation that is tender and precarious and precious. In contrast, the ancient Greeks—like many peoples outside the Abrahamic traditions—took it as indisputable that the *kosmos* was eternal, that it had never emerged and arisen. Why, indeed, would this sense of cosmic dynamism—of waves bursting forth from non-being as from a womb, of the constant tension and interplay between darkness and light, or between the plenitude of the divine countenance [*prosōpon*] and its troubling withdrawal—why would this radically emergent sense of the natural be emphasized by a people who thought the *kosmos* to be eternal?

Of course, *physis* is not just emergence as such, but self-emergence, blossoming forth from itself. Indeed, Heidegger maintains that it is being itself rising up from its own concealed abundance, showing itself in these phenomena—and not simply the self-emergent phenomena themselves—that constitutes *physis*. *Physis* is not, he emphasizes, "synonymous with these phenomena, which today we regard as part of 'nature' [but rather it is] *being itself* [*das Sein*], by virtue of which" these things become manifest.[19] That is, *physis*, like *logos*, is a *Grundwort* only because it is a name for being, which is itself the power that manifests itself within emergence and arising. The sense of being that Heidegger wants to elucidate is neither the frozen eternity of Parmenides nor even the persistent continuity and identity woven into change that Heraclitus seeks to name, but something dramatic and dynamic and vital—indeed something truly living, and thus something better named by the word *ktisis* than by the word *physis*. But this was gradually obscured by the Latin notion of *creatio* that still reverberates in our present-day word *creation*.

III

In the course of the Latin Middle Ages, coordinate to the concept of *creatio*, God becomes understood as "first efficient cause." And if *ktisis*—the radical founding or instituting through saying, the pure performative "let there be," whose Hebrew counterpart *bara* is used in the Old Testament only of God in his act of creation—is translated as *creatio* in the sense of producing (corresponding, instead, to the Greek *demiourgein*) and production is then understood as efficient causation; and if *ousia* is grasped as *substantia*, the self-subsisting, then the path to the modern concept of nature becomes clear enough. But it seems clear, too, that only by deploying this schema anachronistically can Heidegger dismissively state, "the Christian idea of the causation of all beings through a first cause is metaphysical, especially the version of the creation story of the Old Testament as rehearsed in Greek metaphysics."[20] But it may be less clear that this path to the modern concept and experience of creation as nature entails the regression to a fundamentally idolatrous relationship, dramatically obscuring the earlier, iconic character of creation.

For the approach to creation as *ktisis*, in both scripture and patristic thought, always proceeded iconically, seeing creation as an immanent window or icon, opening upon and revealing the transcendent creator. That is, the movement here from creation to creator is *not a movement of inductive reasoning*—neither the inductive generalization of causal inference nor the reasoning of analogical induction—but rather something aesthetic and intuitive and noetic. In the language of Plato's "divided line" it is the "double seeing" of *eikasia*, authentic imagination, not the single vision of *phantasia*, mere fantasy. That is, it is a matter of *seeing through* the image to the original, of listening to the word within the sonority of the phonemes, an act of attending to the transcendent within the immanent. Moreover, this movement from creation to creator is inherently relational and personal, following from the discernment that what is expressed and revealed in creation is meant for me, intended for me, *faces me*—letting it engage and address me, drawing toward the countenance that manifests itself within it.

But prior to this, as with every icon, creation too must first be approached with humility and reverence, with a prior *metanoia*, if I may employ the Greek word for the symbolic prostration that is always made before the icon is venerated with a kiss. And thus what

St. Isaac of Ninevah calls "the glory of God that is hidden in creation" can be revealed only through prior *askesis* and consequent *kenosis* or self-emptying, a self-emptying that corresponds to the *kenosis* through which the eternal Logos creates the world and enters into it through the *logoi* of creation. Incarnate within the visible things of creation, suggests St. Maximus the Confessor, "the Logos both hides himself in his self-revelation and reveals himself in his hiding of himself."[21] (Heraclitus, too, we recall, had commended a certain purification, or *askesis*, for the Logos hidden in things to become manifest.)

How radically different from this is the reified, prosaic, "naturalistic," ultimately idolatrous concept of the *kosmos* as self-contained substance that came to prevail in the Latin West—a self-subsisting, autonomous reality with no inner relation to anything beyond it. Nature is merely an object that is just there, *vorhanden*, but from which we can nevertheless infer another absent, external substance as its cause. But as St. Athanasius saw clearly in the fourth century, the freestanding appearance of the idol conceals its constitution not in divine *ktisis*, or in abstract reasoning, or even in an over-zealous love for creation, but rather in the dialectic of human desire.

Idolatry, Athanasius argues in *Against the Pagans*, derives not from loving creation too much, but rather from turning *away* from it. The Fall, he maintains, is not a fall *into* nature, but rather a fall *away from* it and into ourselves. Athanasius argues that creation naturally draws and entices and enraptures us in its primordial, iconic transparency toward a transcendent creator; yet we turn instead toward what is closer to us, toward what is our own (*idios*)—indeed, toward and into what *is* merely ourselves. And since what seems closest, most our own, is the body, the fall into ourselves becomes a fall into the flesh and the merely, mutely, sensual—a fall into what William Blake would later call "single vision," a seeing that fails to "see through" the *eikon* or image to the original. But Athanasius is by no means denigrating either the visible or the bodily as such. On the contrary, the *visible* is understood as an ongoing gift of the divine presence. Glossing the prophet Isaiah ("Holy, holy, holy . . . the whole earth is full of His glory") Athanasius emphasizes that "every part of creation manifests the glory of God."[22] And it is by means of the *body* that we are embedded in and inhabit this beautiful order (*taxis*) of creation—it is the vehicle through which the *nous* (that is, our open and attentive awareness) can experience it and contemplate it and enter into a relation with the creator *by means of* the natural world—an intermediary through which the visible world may be joined together with the invisible.

Preferring what is near, staying close to shore, and preferring our own—affirming the same at the expense of the other—we fall into ourselves. And this means that we fall into the body as into a prison or cave, as into a pit. Or rather, by falling into our body and away from its naturally ecstatic relation toward the other, it *becomes* a cave and a pit—becomes perhaps uncannily similar to what is valorized by Gilles Deleuze as a "body without organs." And here, alone within ourselves, we discover pleasure—more precisely, the desire for pleasure for its own sake—as a counterfeit of the soul's inherent, ecstatic urge toward transcendence. For, as with Plato's account of *eros* and Husserl's theory of intentionality, Athanasius sees motion as the inner principle of the soul. So when the soul's motion is not directed toward the transcendent creator who addresses it through the beauty of visible creation—toward which it inclines in the manner of a lover responding to the call of the beloved—it moves instead to seek pleasure itself as its new (counterfeit) object, exchanging joy for pleasure, ecstasy for mere immanence, eternity for transience. Absorbed now in itself, the soul nevertheless senses the need to find a terminus for its desiring, and it creates for itself fictions, phantasms, simulacra, *idols* that serve as loci of desire.

And so it is that idolatry is born. "For human beings," argues Athanasius, "having fallen into the unreasonableness of their passions and pleasures, and unable to see anything beyond pleasures and lusts of the flesh, inasmuch as they keep their mind [*nous*] in the midst of these irrational things, they imagined the divine principle to be in irrational things [themselves], and carved a number of gods to match the variety of their passions" (*Contra Gentes*, 19:1).

But since this turn away from God and that which is (*to onta*) is a turn toward nothingness (*to mē on*) and therefore dissolution, Athanasius argues that these fictions or simulacra become more numerous and fragmented—more incoherent and dissonant, more debased—with the worship of animals such as sheep and oxen and crocodiles, and even inanimate objects—and therefore more demeaning to those who worship them. Eventually, these idols turn against the people who made them and worship them, demanding they sacrifice one another to the fictions they have themselves fabricated. A sad irony, for these same elements and creations can be mistaken for divinities only because in their very being, they *already are* manifestations of the magnificence and glory of God.[23]

The descent into idolatry, turning away from *what is* into what is not, falls into dissolution and chaos, while the idols become fearsome, objects of an unholy, and indeed demonic, enchantment—sources of dread. Veneration then takes the form of placation and propitiation and anxious appeasement. As the Russian philosopher Pavel Florenksy argued, "Antiquity did not know the enjoyment of nature." For the love of nature, the "sense for nature" as we have come to know and embrace it, presupposes the perception of nature as *ktisis*, creation, addressing and embracing us with divine energies—as the initial and ongoing revelation of a loving and living God, whose Word to us *is* creation itself. In contrast, Florensky argues that "shackled with fear, ancient [man directed] all his energies at the 'skin' of things and at the reproduction of this 'skin.'" "The character of ancient art," he continues, with perhaps a sly nod toward Nietzsche's praise of surfaces, "shows that ancient man did not at all love the 'soul' of things and was afraid of getting beneath the 'skin,' for there he found chaos and terror. . . . 'Better not to look,'— that is the slogan of an ancient culture that tried to forget itself in 'optimism,' the optimism of an opium-eater or hashish-smoker. Here, only a formal science is possible; geometry, astronomy . . . and so on."[24] The dependable regularities of nature that Hesiod invoked through libation and sacrifice become the calculable objects of geometry. But as the interiority of nature is shunned in favor of its surfaces, its very reflectivity can now radiate the glimmer of our own desires, a possibility that is seized upon cultically in the rites that Heraclitus found shameful and much later in the technology of nature that governs the modern world. The solid geometry of substantial nature supports the radiance of its borrowed light, reflecting back to us our own desires. The window becomes a mirror; the icon becomes an idol.

IV

"The world is all that is the case," begins Wittgenstein in his *Tractatus*; "the world is the totality of facts."

"The heavens declare the glory of God, and the firmament proclaimeth the works of his hands." So begins the 18th Psalm of David, in the Septuagint.

But consider also the saying of St. Nikitas Stithatos, an eleventh-century Byzantine monk, compiled in the *Philokalia*: "the nature [*physis*] of things changes according to the inner disposition of the soul."

"The nature of things changes"—*physis* manifests itself differently, says St. Nikitas, "according to the inner disposition of the soul."[25] Moreover, he continues, when the soul is disordered, then perception no longer "operate[s] in a way that accords with nature," and the "natural beauty" of things "does not exalt [the soul] to an understanding of their Maker."[26] But to those whose souls are ordered and have attained a certain clarity of spiritual vision (*theōria*)—to those like St. Anthony the Great, the first ascetic to enter the desert wilderness in search of the Logos—all creation manifests the *logoi* of its creator: "My book is this created nature," explains St Anthony; "it is always with me, and whenever I wish I can read in it the words of God." Or, as put by St. Gregory of Sinai, one of the first great expositors of hesychast asceticism, "he who has been elevated toward God sees . . . by the grace of the Holy Spirit, all of creation as *luminous*."[27]

But it is otherwise to the idolatrous eye, which outside the spheres of its peculiar idols and fetishes sees nature generally as opaque and

flat and impervious. Here we might indeed say, along with Max Weber, that the world has lost its enchantment; it has become an indifferent aggregate of facts, or as Wittgenstein puts it, "all that is the
case." And ultimately, as Heidegger maintains, not just a mute presence, something *vorhanden*, but a consumable stock or inventory:
Bestand. And surely this can happen to an entire society, indeed an
entire epoch, just as much as to an individual person. Dostoevsky
was deeply cognizant of this basic spiritual principle. In his later
novels, he repeatedly shows how spiritual recovery results in a new
way of experiencing nature and other people, an experience of nature
as everywhere gleaming with divine energies, and of other people as
profound and wonderful and mysterious, as imaging the God who
created them. And conversely, when the soul is disordered—as in
the case of Raskolnikov, who finds both elemental nature and individual people to be oppressive until his soul finally begins to heal;
or the Underground Man, whose spite extends not just to his fellow
humanity, but to the very order of creation itself; or Ivan Karamazov
who explicitly rejects that order—then nature and humanity alike
withdraw their luster, becoming dull and vapid, or else demonic and
menacing.

Florensky, too, argues that it is only through genuine asceticism
(as opposed to the false asceticism that despises creation) that the
beauty of nature can be revealed—that it becomes possible "to live
and feel together with all creation, not with the creation that man
has corrupted but with the creation that came out of the hands of
its creator"—that is, through a perfected correlation between what
St. Ambrose called the "inviolate and undamaged nature" of the
soul and of creation itself.[28] Or, as we find it expressed in nature
writing from Thoreau and Muir to Annie Dillard and Wendell Berry,
the holiness of creation can be encountered only through a kind of
purification of the soul, an *askesis* from what Muir saw as the toxic
fumes of society and desire and consumption. It is fitting, then, that
it is within the discourses of the monk and ascetic Zosima that Dostoevsky presents a powerful expression of this sense of creation as

revealing the face and gift and greeting of God—this sense of nature that I believe Heidegger mistakenly reads back into the Greek word *physis*, overlooking its roots in creation as *ktisis*.

In *The Brothers Karamazov*, Dostoevsky's last great novel, the Elder Zosima narrates two remarkable conversions, two changes of heart, each of which shows an intertwining of three elements: repentance for wronging others; an embracing of ascetic renunciation; and a dramatically new experience of nature as creation. In the first story, Zosima's young brother Markel, fatally ill from tuberculosis, realizes he has been treating his mother cruelly. To atone, he decides to please her by beginning belatedly—on Tuesday of Holy Week—to keep the Lenten fast he had previously scorned in unbelief. And this heralds a change in his demeanor: from his usual mode of irritability and ill-temper, he becomes generous and joyous, and begins to see creation in a whole new light—indeed, in a "luminous" manner. He even apologizes to the birds outside his window for his lifelong unpleasantness: "Birds of God, joyful birds, you too must forgive me, because I have sinned against you. . . . There was so much of God's glory around me: birds, meadows, sky and I alone lived in shame, I alone dishonored everything, and did not notice the beauty and glory of it all." He goes on to articulate one of Doetoevsky's last, great themes, by exclaiming that we are already, now, living in paradise—that is, living in a world where all creation is transparent to its creator—yet we are strangely unable to see it, "we do not want to know it."[29]

Later, as a young cadet, on the eve of a duel that he has provoked, Zosima himself cruelly slaps his servant. Awakening the next morning, he looks out upon a lovely scene: "the sun was shining, the leaves were rejoicing, glistening, and the birds, the birds were praising God." Suddenly remembering his late brother Markel, he weeps in repentance, and asks forgiveness of his amazed servant. But his renunciation is still not complete. He must endure humiliation by showing up at the duel with no intent of fighting. Allowing his opponent to take the first shot, he flings his gun into the woods, and exclaims to the shocked witnesses his new awareness of creation as a gift: "'Gentlemen, look at the divine gifts around us: the clear sky, the fresh air, the tender grass, the birds, nature is beautiful and sinless, and only we, we alone are godless and foolish, and do not understand that life is paradise, for we need only wish to understand it, and it will come at once in all its beauty, and we shall embrace one another and weep."[30]

Nor does this new experience of nature as *ktisis* or creation diminish. Zosima relates a more recent story when, traveling through Russia, he spent the night on a riverbank with some fishermen and a certain young man who, despite his harsh life pulling barges with a rope, looks upon the world "with a clear and tender gaze." "It was a bright, still, warm July night," Zosima narrates, "the river was wide, a refreshing mist rose from it, once in a while a fish would splash softly, the birds fell silent, all was quiet, gracious, all praying to God. And only the two of us . . . were still awake, and we got to talking about the beauty of this world of God's, and about its great mystery. For each blade of grass, each little bug, ant, golden bee, knows its way amazingly; being without reason, they witness to the divine mystery, they enact it. And I could see that the lad's heart was burning. He told me how he loved the forest and the forest birds." And Zosima replies, "How could it be otherwise . . . for *the Word* is in all, all creation and all creatures, every little leaf is striving toward the Word, sings glory to God, weeps to Christ, unbeknownst to itself, doing so through the mystery of its sinless life."[31]

I want to suggest that this vision of the natural world—the mist rising gently from the river, the fish splashing softly in the night, the divine mystery quietly emerging in all things—this experience of the expressive Word inhabiting the very pith of things shows the true locus of nature as self-emergence. It engages nature as *physis* in Heidegger's sense, but more properly articulates nature as creation, as *ktisis*. It is deeply resonant of a host of passages from the Psalms, from the Book of Job, from the prophets, and from the Sermon on the Mount. And it is a world away from Greek literature and philosophy.

V

This experience of *ktisis* is articulated in patristic literature by means of a range of teachings, four of which it will be helpful to characterize more systematically. First, the earth and materiality are affirmed to a degree unprecedented in Greek philosophy. Material creation is embraced not just as the first, and ongoing, revelation of God, but as the very element to which God will ultimately be joined—as that of which St. John of Damascus, in his defense of the veneration of icons, is able to affirm, "I do not venerate matter [as such], I venerate the fashioner of matter, who became matter for my sake and accepted to *dwell in matter* and through matter worked my salvation, and *I will not cease from reverencing matter*, through which my salvation was worked. . . . Therefore *I reverence the rest of matter* and hold in respect that through which my salvation came, because it is *filled with divine energy and grace*."[32] Contrast this with a commentary by the neo-Platonist Amelius, Roman student of Plotinus, on the Prologue to the Gospel of St. John, to whom he refers condescendingly as "the Barbarian":

> This, then, was the *Logos*, thanks to which all engendered things were produced while he himself exists always (as Heraclitus believed) and which the Barbarian believed "was near to God" and "was God," possessing the rank and dignity of a principle. . . . And it fell into bodies and, donning flesh, [it] took on human appearance; but at the same time it shows the greatness of its nature. When it is freed, it is once again made divine, and is God, [just] as it was before it fell into the world of bodies, and before it descended into flesh and humankind.[33]

The Logos here is not a personal reality, a "he" or "whom," but a principle, an "it" and a "that." Thus, the Logos only appears to be human, having undergone a "fall" into matter and particularity, from which it is finally "freed," allowing it to "once again" become divine. Materiality too is utterly incompatible with divinity here.

Nor is this an outlying view; it continues and culminates in the understanding of matter and materiality prevalent throughout ancient Greek philosophy.

Second, taking the Incarnation of the eternal Logos not as something strange and anomalous, but as a completion and fulfillment of creation itself, patristic thought understands the Second Person of the Trinity to be always already embedded in creation, for it is through him that all things are made. This proto-incarnation, as it were, is articulated through the concept of the *logoi* of creation: inner, inexhaustible sources of meaning and intelligibility that inhere in all things and that allow them to be what they are—the interiority *from* which, to use the language of Heidegger, they come forth in their self-emergence—with each of these *logoi* reflecting, in a unique and unrepeatable manner, the eternal Logos. The *logoi* of creation are what Evagrius of Pontus in the third century called "the divine Logos in the ontological heart of things."

Third, these inner principles of creation—the true *physeis* of each thing, every "blade of grass, each little bug, ant, golden bee"—can be revealed only to the eye of the heart, to a vision that is emptied of selfishness and desire, to the soul restored to order though *askesis*. Thus, since Evagrius, *askesis* has been understood to prepare the soul for what the Desert Fathers called *theoria physike*: the seeing of the inner meanings of creation, the illumination that grants us access to the self-emergence of creation. *Theoria physike*, rather prosaically translated as the "contemplation of nature," itself opens up the possibility of living Edenically, within a restored and renewed creation, seeing and tasting in this world the resurrection and restoration and transfiguration of all things. At the same time, however, it must be emphasized that just as the Logos here is not some kind of impersonal process of ontological unfolding, as Heidegger implies, but a living and personal reality—and therefore, the particular *logoi* must be seen not as neutral effects of an indifferent cause, but expressions just as personal as the paintings of Cézanne—so the exercise of *theoria physike* is a personal relation. As argued by the contemporary Greek philosopher Christos Yannaras, by means of its beauty and the wonder it generates, creation presents itself as call or invitation (*klesis*) toward a relation to that personal reality from which the call itself issues, and to which it refers. The wonder-laden passages of the Psalms and other Wisdom Books of the Bible, where the creator is apprehended and praised within the experience of creation, exemplify this kind of response, which moves toward restoring humanity to its true vocation—serving as the proper priesthood of creation, exercising its ancient ordination of seeing the invisible within the visible, seeing all things in God, and joining together heaven and earth in a cosmic consecration. It is toward this great end that the experience of *ktisis* always strains and leans.

Finally, how is this inherence of God in creation possible without collapsing transcendence into immanence—without some kind of pantheism? Patristic thought makes a crucial distinction between the divine essence or *ousia*—which is radically transcendent, always mysterious, and forever unknowable even to the highest angelic orders—and the divine *energeiai* (energies or activities) that, far from being transcendent, surround us everywhere, revealing themselves within and through the *logoi*, but also within our own hearts. Rooted in the thought and experience of the Cappadocian and Alexandrian theologians and the Desert Fathers and Mothers of Egypt and Syria and Palestine, developed richly within the teachings of St. Maximus the Confessor, and culminating in the work of the fourteenth-century Byzantine monk St. Gregory Palamas, this vitally important distinction of essence and energies was poorly understood and eventually rejected in the West, which as a result reverted to positing an increasingly transcendent deity, in relation to whom immanence becomes more and more incomprehensible. As a result, God is increasingly approached not through direct experience, but through the discursive rationality that conceptualizes the relation of God to creation as one of cause and effect and no longer through the immanence of divine energies at work in the world. As Yannaras argues, rather than approach God through a personal and essentially erotic response to the invitation that inheres in all natural beauty, Western modernity infers and extrapolates the creator's existence based upon the analysis of creation, seeking to think its way back to God. A relational ontology is displaced by a rationalistic metaphysic. Or so this development can be seen from the perspective of the Byzantine East, which most faithfully preserves the essential experience of *ktisis* and its relation to the creator in patristic Christianity.

VI

But even if certain Eastern elements of this understanding of *ktisis* faithfully preserve the tradition that leads from the poetics of Genesis and the Wisdom Books to the insights of the great desert-dwelling mystics of the patristic era, shouldn't the *experience* that corresponds to this understanding be manifest and discernible in the Christian West as well, whatever its degree of errancy may or may not have been? And, indeed, what are the main features of creation as a phenomenon, as it is encountered in experience, that might constitute the initial steps toward an anatomy, taxonomy, or indeed phenomenology of creation?

First, we may look to the seventeenth-century Anglican priest and poet Thomas Traherne, who celebrates the *gift-character* of creation. Traherne's great, poetic insight is that in the natural world around us—in simple things like sunlight and air and the earth at our feet—are gifts of incomparable beauty and inestimable value—gifts that we are not just entitled, but *obligated* by our very constitution, to regard as given to each of us alone, paradoxically making us each the sole "Possessor of the Whole World."[34] Teaching "Things Strange yet Common; Incredible, yet Known," he asks, "Is it not a Great Thing, that you should be Heir of the World?"[35] "The Heavens and the Earth Minister unto me," he writes, "as if no Man were Created but I alone."[36] Yet we fail to claim this "Sacred gift," even though it would require nothing more of us than rightly enjoying it.[37] We are in fact inclined to love, made to love; we are secretly drawn toward some great love like "Iron at a Distance is drawn by a Lodestone"— "there is in us a World of Love to somewhat, tho we know not what

in the World that should be."[38] Yet, in confusion, we have settled
for things manufactured, of merely conventional value, falling into
what is nothing less than a "Blindness and Ingratitud, and Damned
folly."[39] Traherne, then, takes as his poetic task the stirring of this
eros, awakening us toward what he calls "those Truths you Love,
without Knowing them."[40]   .

> "Your Enjoyment of the World is never right," he urges, "till
> evry Morning you awake in Heaven . . . and look upon the
> Skies and the Earth and the Air, as Celestial Joys: having such
> a Reverend Esteem of all, as if you were among the Angels."
> (1:28) "You never Enjoy the World aright, till the Sea it self
> floweth in your Veins, till you are Clothed with the Heavens,
> and Crowned with the Stars; and Perceiv your self to be the
> Sole Heir of the whole World: and more then so, because Men
> are in it who are evry one Sole Heirs, as well as you."[41]

That is, Traherne shows that seen from within the *askesis* of a
life freed from desire, accumulation, and obsessive consumption,
the earth reveals a transitive and indeed emergent character: it is
always just being given! (And we note in passing that Traherne presents poetic evidence counter to Derrida's claim that the gift is not
possible as a *phenomenon*.) But it is even more important to notice
that, as with any genuine gift, this is not an impersonal process—
indeed, here it entails the generosity of a donor set upon lavishing
the goodness and beauty that he is.

Second, creation is not itself impersonal, but presents to us a face.
That is, as was argued previously, creation is an icon of God, the first
icon offered up for the veneration of all humanity. John Muir, for
example, boldly proclaims that "God's love is manifest in the landscape *as in a face*."[42] And the writings of the deeply pious Muir—
especially as found in his journals and correspondence, before timid
editors had deleted and secularized his countless references to God
in nature—constitute perhaps the most impressive literary testi-
mony to the experience of nature as presenting a living face—the

smiling, loving face of its creator. Muir's younger Russian contemporary, Fr. Pavel Florensky, also writes of the face of nature in his discussion of iconography, remarking that "we can speak of all natural things and creations with whom we are in conscious relation as having a face: as, for example, we speak of the *face of nature.*"[43] Just as much in love with nature as Muir, Florensky too looked upon the face of nature as if venerating a written icon.

Unlike Marion, who distinguishes only between face and façade, Florensky distinguishes face, mask, and countenance. The latter is of special interest here. Whereas for Florensky the *face* presents a dynamic, shifting boundary between an outward aspect and a dynamic interiority that "draws us in" and absorbs our consciousness, the *countenance* is the unimpeded shining-forth *of* this very interiority—that is, its energy or actuality as it fully animates the face, transfiguring it into material reality that is transparent to spirit, manifesting "the mysteries of the invisible world" and actualizing "the deepest tasks of its own essence."[44] It is "revealed spiritual being," "eternal meaning contemplatively apprehended," the "appearance to us" of what Plato called *eidos* or "inner look." Using the language of image and likeness from the Genesis narrative, the countenance must be understood not just as the incipient *image* of God, taken as an inner potentiality, but as its shining out with such radiance that *through the countenance*, the face is now being transformed into a divine *likeness.* "Everything accidental" is "swept away" by the "energy of the image of God," and nothing is visible but the divine radiance incarnate in the face.[45] That is, the countenance that shines through the face of creation is itself the glory (*doxa*) of its creator. And here again, we note what Heidegger ignores: that the relation to this dynamic interplay in the natural world is a personal one, not an indifferent attention, as to a process that would leave one untouched and unaddressed.

The experience of being addressed by creation, and especially by its beauty, manifest in the works of countless nature writers during the last two hundred years, is taken up in Yannaras's understanding of *prosōpon*: face, countenance, or presence. As Yannaras maintains, our wonder or surprise "before the beauty of the world" is "like the surprise before the incomparable and unique beauty of every offering (*prosphora*) of *eros*, which transforms the *gift* into an *icon* of the beloved who has given it."[46] And we recognize the giver within the gift in much the same way that we recognize van Gogh in one of his paintings—even for a new painting that I have never seen, it is entirely possible to say, "yes, *that is* van Gogh; I would recognize him anywhere!" Creation for Yannaras, then, presents us with a call, summons, or invitation (*klesis*) to relation through its beauty. But once again, it is the call to a personal relation, not to an impersonal witnessing or heeding. Finally, as do Traherne, Muir, and Florensky, Yannaras also emphasizes the need for *askesis*, a purification of the soul, to enter into this relation, to stand in this presence. And we may wonder here whether Heidegger's *Gelassenheit*—"calm composure," letting-be, or willing non-willing—is spiritually comparable to traditional *askesis* and the *metanoia* or "change of heart" that must accompany it, or whether it is instead a chilly, abstract, and truncated analogue.

VII

Finally, what of the dyadic character of *physis* in Heidegger's exposition: the sense of *physis* as both a self-emergence and a self-withdrawal? From whence, and to whence, does this arising or emergence proceed, if we are to better understand this distinguishing character of creation as *ktisis*, rather than *physis*? It is the divine *logoi* from which all creation emerges, and back into which it recedes—and that are ever manifest and concealed according to our capacity to see them. Nor are these Platonic forms or Augustinian exemplars, but the inexhaustible meanings, willed by the creator, that inhere in all things—the divine energies themselves, which are always emerging as *energeiai* or activities of the ever-mysterious divine essence. At the same time, however, they are manifestations of the divine glory (Greek *doxa*, Hebrew *shekinah*, Arabic *sakinah*). And as the twentieth-century Russian philosopher Sergei Bulgakov maintains, "in the light of this Glory we can discern the glorified Face of Logos-Christ."[47] In scripture—for example, in the theophanies on Sinai—this glory of God, as Bulgakov notes, always entails a dyadic alternation of light and darkness, especially of dark cloud and inner lightning bursting forth. And so it is in the narratives of the Transfiguration, when in the midst of the dazzling light of Christ a cloud descends, and, as it is said in Luke, actually envelops the anxious apostles.

Yet it is not only Christ who is transfigured on Mt. Tabor, but his garments, as well, which are described as "shining, exceedingly white, like snow," and as becoming "as white as the light" itself. Thus we find revealed here not just the glory of God as such, but the glory of God *in creation* as well, in the very raiments of God incarnate. Bulgakov comments, "The metaphysical structure of the world is changed or rather revealed by the Transfiguration, without any change in the physical nature of the world, whose foundation was unalterably laid during the Six Days of creation." The world, he continues, "is clothed in beauty."[48] Nor is this an imported or borrowed glory, but the shining forth of the logos-structure within which it has always been rooted, and that the apostles are momentarily allowed to glimpse. It is this same glory of God in creation— this holy glory of God that Isaiah tells us fills the earth—that Hopkins describes in saying that "the world is charged with the grandeur of God, it will flame out like shining from shook foil," as it emerges from "the dearest freshness deep down things."

The inner *logoi* of creation, then, would in relation to us constitute a revealing of God—a gift, a summons to relation, and indeed a countenance—but in relation to creation itself, they would constitute its genuine interiority, the "self" from which self-emergence proceeds—in contrast to the darkness of earth that Heidegger suggests as the receding pole of *physis* and that strongly resembles the recalcitrant, dark matter that found expression in Gnosticism and ancient dualism. But traditional Christianity is by no means silent about the darkness of earth that quietly, gracefully, humbly harbors the light. Readers of Dostoevsky's later novels will be familiar with evocations of the dark and moist earth in passages such as this prophecy from a strange and mysterious old woman: "The Mother of God is our great mother the moist earth, and therein lies a great joy for man."[49] We recall, once again, the prophecy from Isaiah: "the whole earth is full of His glory." In line with the Orthodox tradition, Bulgakov understands this as a prophecy of the Mother of God: "she is that earth which is the receptacle of God's Glory, the fulfilled Glory of the world." And as such the Mother of God, and with her the entire earth, is filled with light—the light of transfiguration that Christ allowed the three apostles to experience.[50]

Thus, with the divine glory in creation, we arrive at the two central icons situated at either side of the royal doors at the center of the iconostasis, through which heaven and earth interact and mingle and commune during the divine liturgy of the Orthodox Church. On the right side is Christ as *Pantocrator*, both *arche* and *archon* of all. And on the left is the Mother of God, upholding the infant Christ. Creation must ultimately be understood as iconic both of Christ the Logos and of his Holy Mother who bears his Incarnation—in each case addressing us with a radical call to relation, even while summoning us to repentance and *askesis*, and thus calling us toward love, as well as toward transfiguration—our own and that of creation itself.

Creativity as Call to Care
for Creation? John Zizioulas
and Jean-Louis Chrétien

Christina M. Gschwandtner

Thus is the high priest arrayed when he sets forth to his holy duties,
in order that when he enters to offer the ancestral prayers and sac-
rifices there may enter with him the whole universe, as signified in
the types of it which he brings upon his person. . . .
For, as he wears a vesture which represents the world, his first duty
is to carry the pattern enshrined in his heart, and so be in a sense
transformed from a man into the nature of the world; and, if one
may dare say so . . . be himself a little world, a microcosm.
                                        —*Philo of Alexandria*

Jean-Louis Chrétien, employing this quote by Philo and drawing on
various aspects of the Christian tradition, argues for a "cosmic liturgy" centered in the human being who "can in his song, offer the
world to God."[1] Humans are able to provide an ark for all creatures
as they offer them to the divine through their creativity and via what
Chrétien calls a "eucharist of speech." Offering the creation to God
in praise and hymnody is both a special task for human beings and a
response to the divine call. Humans hence carry a responsibility for
the world: they shelter, protect, and shape it. At first glance, such
a vision of care for creation in terms of human creativity and based

on a divine call seems eminently useful for promoting ecological concerns for creation. Indeed, this is precisely how it is articulated by various Eastern Orthodox theologians, who—one would think— could not be further removed from contemporary French phenomenology. Yet, the unique role they see human beings playing on behalf of creation is strikingly similar: Humans are "priests" of creation who have a Eucharistic task to carry all of creation in praise to God. And this is explicitly understood by Eastern theologians as a kind of "cosmic liturgy" that provides justification for an ecological vision. In fact, many Orthodox thinkers now consider this *the* distinctive contribution Eastern Orthodox theology can make to the ecological debate.[2] Eastern and indeed some Western theologians regard the notion of human priesthood on behalf of creation as far superior to traditional Western notions of stewardship for or "kingship" over creation, ideas that have been rejected by many as providing license for exploitation because they posit human beings as distinct from other living beings and conceive of them as rulers over all of creation (even if only as "stewards" for the divine).[3] This predominantly Western idea of human kingship or stewardship is rooted in the verses in Genesis that speak of the human as being made "in the image and likeness" of God and given the task to work the ground and name and rule the other creatures.[4] Humans, therefore, are superior to all other creatures, the apex of creation, the ones for whom everything else was made. They are special and unique: Creation provides the stage on which humans are the only real actors in the drama of redemption.[5] This exclusive focus on the human, which separates us from the entire rest of creation and disregards biological and ecological insights about our connectedness to and dependence on other creatures, has been condemned in recent years as at the very least unhelpful for greater care for creation or possibly even the cause for ecological destruction.[6] Yet, is the Eastern notion of priesthood really so very different?

The ancient idea of the *imago dei* plays a significant role in the Eastern tradition.[7] There also, it somehow designates that humans are unique, although there is no full agreement about what exactly the image means.[8] While it was often identified with logos, this was by no means always the case, and it was interpreted in terms of upright posture, rationality, morality, sexual differentiation, tool-making, self-consciousness, or even simply having a soul.[9] Regardless of how it was defined, the *imago dei* was generally employed as a way to express human uniqueness and human distinction from the animal world.[10] For many thinkers this was linked with the conception of the human as microcosm, articulated by Plato and Philo: humans are a "little world" in that they encapsulate the entire universe in themselves, consisting of matter, animality, spirit, and soul; they are on the boundary line between physical and spiritual, thus both animal and spirit.[11] They stand on the dividing line between the material and the spiritual spheres and participate in both.[12] They are the knot that holds all of creation together, and they have special responsibility for it. While many Patristic thinkers refer to this idea of the microcosm on some level, Maximus the Confessor articulates it the most fully by assigning to the human being the task of overcoming all the major divisions of reality and unifying them in redemptive fashion.[13] The contemporary concept of a human ecological priesthood usually relies heavily on Maximus as its main source. Here I will review and evaluate the version this idea takes in one contemporary Orthodox thinker who has employed it with the most nuance and has argued most forcefully for it as an important contribution to ecological discourse: John Zizioulas, who has for years functioned as close advisor to the "green" Patriarch Bartholomew. I will then return to Chrétien in the second part of my discussion.[14]

John Zizioulas: Human Creativity and Care for Creation

Zizioulas admits that humans are animals practically indistinguishable from chimpanzees.[15] He also recognizes that most of the traditional ways of phrasing human uniqueness have fallen short, in that subsequent research has shown that some animals also display these traits to a greater or lesser degree. Instead, he proposes to articulate the *imago dei* as "freedom in otherness" and ultimately in

terms of "a drive towards love and creativity." This freedom is essentially about creativity; humans are the only ones who can create
alternative worlds, who can project themselves into a work of art
and portray the other. They find uniqueness through this creative
self-expression: "The freedom to be other involves the tendency
to create a world other than the given one, that is, to bring about
otherness in the radically ontological sense of the emergence of new
identities bearing the seal of the lover's or the creator's personhood.
This is expressed in art, when it is not a mere copy of reality, and *it
is a distinctive characteristic of the human beings in creation. . . .
What the human being aspires to achieve through otherness is not
simply difference but uniqueness. This is also a distinctive mark of
humanity related to the *imago Dei*."[16] Thus what makes us unique
as human beings is our drive to creativity, our ability to produce
art.[17] This creativity depends on a special divine call to us, a call
that is not heard in the same fashion by other creatures. This call
establishes our personhood and our capacity for relation, both ideas
central to Zizioulas's theology. The call, as in Lévinas, proceeds
from and maintains otherness, but for Zizioulas (unlike for Lévinas,
on whom he relies here but also criticizes) it can be identified as a
gift from God. And this gift of the call establishes our uniqueness:
"Through the call, Adam is constituted, therefore, as a being other
than God and the rest of creation. This otherness is not the result of
self-affirmation; it is an otherness granted and is not self-existent,
but a particularity which is a gift to the Other. Thus, while the rest
of creation is other than God and other creatures only in the form of
species (God created the plants, the animals, etc., without addressing them with a call), the human being is singled out, not merely as
a species, but as a particular partner in a relationship, as a respon-
dent to a call. This is the constitutive event of humanity. Outside
this event of divine call, humanity is part of the animal species."[18]
Zizioulas argues that this is a new and better way of phrasing the
imago dei,[19] because it no longer employs traditional substance and
nature language and instead expresses human uniqueness in terms
of relationship and activity, not in terms of being.[20] He interprets the

fall as a way of denying this divine call, which reduces Adam to a mere animal and denies his special humanity: sin is a return to bestiality, to making us "mere" animals.[21] Christ recalls fallen Adam and humanity to the original call for free and loving relationship with God and creation.[22] Humanity, then, bears God's "personal seal."[23] Human relationships are patterned on the divine and have a special priestly task that imitates and assumes that exercised by Christ.[24]

Zizioulas contends that human uniqueness leads to a special responsibility for creation that we are to "hypostasize."[25] He explicates this process in terms of love, which extends hypostasis (i.e., unique personhood) to the other.[26] This would seem to imply that our relation with the rest of creation should be one of loving care that makes the other equal to us in some way. Yet it is pretty clear that Zizioulas extends personhood and thus loving relation only to other humans and does not include the rest of creation within it. Indeed, unless he is specifically addressing environmental issues, animals or plants do not usually appear on his theological horizon.[27] The loving hypostasization is ultimately encapsulated in ecclesiology, and it is only at this point that nonhuman creation emerges again via the priestly Eucharistic office.[28] Through the church's hypostasization, all beings participate in eternal life in some fashion. Zizioulas articulates this in terms of a "eucharistic ethos" that has an "ecological significance."[29] This ecological significance consists in our care for creation, which offers creation to God as a gift within the liturgy. Through this it becomes "our creation" and is sanctified:

> A eucharistic way of being involves respect and care for the whole creation. The Eucharist is a "cosmic liturgy" in which the human being acts as the "priest of creation" offering to God with gratitude the gift of created existence as the body of him who freely assumed this existence in his own hypostasis in order to "save" it, that is, to assure and confirm the survival of creation. In giving thanks for creation we do not simply utter words of gratitude to the Creator. We take creation in our hands and offer it to the Creator and to our fellow human be-

ings as our own personal gift, as our own creation. Thus the eucharistic way of being involves an act of dedication or "setting apart," a sacralization of creation.[30]

This implies a special task for humans, which they alone can accomplish: "The human being is called to bring the rest of creation into communion with God, so that the hypostasis of every creature might be saved from mortality and thus be shown to be a true *hypostasis*, that is, truly existing as particular and 'other,' and not swallowed up by the general." The drive of creation for survival "can be fulfilled only in and through the human being." This happens not only in church but also in everyday life as "human beings take creation in their hands in order to liberate it from mortality and lift it up to truly hypostatic existence."[31] Zizioulas sees this confirmed in Byzantine art and its lack of interest in landscape painting—nature is liberated in art.[32] Creation and creativity are linked, insofar as creativity liberates creation by hypostasizing it.

Thus a particular task of care for creation is imagined for humans. This task is said to be essential for the very survival of nonhuman creatures and ultimately for creation's eschatological immortality. In a text specifically concerned with the ecological crisis, he expresses this even more succinctly:

The priest is the one who takes in his hands the world to refer it back to God, and who, in return, brings God's blessing to what he refers to God. Through this act, creation is brought into communion with God himself. This is the essence of priesthood, and *it is only the human being* who can do it, namely, unite the world in his hands, in order to refer it to God, so that it can be united with God and thus saved and fulfilled . . . *this role of the human being, as the priest of creation, is absolutely necessary for creation itself, because without this reference to creation to God the whole created universe will die.* . . . Therefore, the only way to protect the world from its finitude which is inherent in its nature, is to bring it into relation to God. . . .

> This underlines the significance of man as the priest of cre-
> ation, who would unite the world and relate it to God so that
> it may live for ever.[33]

Humans are able to deliver the rest of creation from its finitude by creatively carrying it to God in praise.

This account of the human call of care for creation through creativity raises a variety of problems. First of all, it is far from clear how creativity is supposed to accomplish any real ecological action. How does an art that eschews landscape painting elevate and sanctify the creation? In what sense do the Eucharistic actions of breaking bread and pouring wine somehow constitute a blessing of *all* of creation? How does *consuming* these elements hallow them or anything else in creation? Is "consumption" really the best way to sanctify natural elements? And how do these products of human agriculture help nonhuman creatures in general? How does it prevent us, for example, from cutting down the rain forest or draining a wetland? How does the creativity Zizioulas proposes keep us from polluting and exploiting the environment? Even if these are valid and helpful connections, how exactly they are to accomplish this work of care for creation needs to be articulated much more fully.

More importantly, however, the role the human being occupies in this exposition is deeply troubling. Humans are posited to be at the very center of the universe, the crux that holds together all of reality. Creation only survives through the human and has to be "hypostasized" by humans in order to be able to relate to God. On the one hand, this seems to imply that nonhuman creatures have no direct connection or relationship to God and that they are not loved or enjoyed by God on their own terms.[34] Nonhuman creatures cannot praise God directly, but require human mediation, human speech. Humans must praise on behalf of other creatures. Yet does God really need our voice in order to sustain a relationship with other creatures?[35] On the other hand, it is again far from clear what "hypostasing" or "humanizing" the creation actually means in terms of practical action. In Sergei Bulgakov such hypostatization amounts to agricultural labor and domestication of animals.[36] In other Orthodox thinkers, such as Staniloae and Evdokimov, it results in a rather unqualified embrace of economic and technological progress.[37] What guarantees that this creative "hypostasization" of creation will be benign or helpful to other creatures? Most attempts at "humanizing" the rest of creation have led to large-scale destruction of habitats. Domestication and pet ownership are not exactly the most environmentally sensitive practices, nor is it clear how these somehow "redeem" nonhuman creatures. When one examines the eschatology of these thinkers one quickly realizes that no real redemption of nonhuman creatures is intended. The creation is redeemed only through human immortality and theosis. While many contemporary Orthodox thinkers such as Patriarch Bartholomew or indeed John Zizioulas do indeed care deeply about environmental issues, they do so only in a rather anthropocentric fashion.[38] And in some other Orthodox thinkers this is pushed even further: the rest of creation becomes entirely dispensable, and humans are seen as divine *by nature*. Not only are other creatures forgotten, but our own creatureliness is erased. While the emphasis on care for creation and the focus on ecology more generally are laudable, this does not seem the best way to articulate it. French phenomenologist Jean-Louis Chrétien may present an alternative that preserves many of the same themes of call and creativity, but articulates them somewhat differently.[39]

Jean-Louis Chrétien: Divine Creativity, Human Creatureliness, and Care for Creation

Chrétien's philosophy is one of hospitality to the voice. This is maybe articulated most fully in his *The Ark of Speech*, where he argues that the human voice can become a sheltering ark for the rest of creation: "Hospitality is, first and foremost, the hospitality that we give each other, exchanging words and silences, glances and voices. And yet, this conversation cannot take place in a vacuum; it is in the world, this world that we never cease to share—among

ourselves but also with other forms of presence, the presence of animals, of vegetables, of things. In our speech, we are equally responsible for them, just as we respond to them."[40] He suggests that we
have a responsibility to shelter other beings in our voice, which he
illustrates with the biblical story of Adam's naming of the animals
and that of Noah's ark, which shelters the animals from the flood.[41]
We must both summon other creatures by our speech and respond
to them. The "silence of the world is broken in our voice" as we
"offer the world in offering ourselves."[42] This "making of a world"
for the rest of creation is "the first and last moral task" for human
beings.[43]

For Chrétien, it is a philosophical task of gratitude and of listening
that is rooted in the flesh and calls forth a response. This response
is especially worked out through beauty and creativity. Beauty calls
us, moves us, touches us.[44] Using Lévinas's notion of the à-Dieu,
he suggests that "beauty says à-Dieu" by offering the things of the
world and by being offered to God.[45] Beauty points us toward God. In
beauty we respond to God's call. Beauty is an excess that overflows
from the divine, and thus in our response to God through beauty we
draw on its prior source in God's abundant excess. Through beauty
we shelter the world. It calls forth our responsibility for the world.
By responding to the beauty of the world, we speak on behalf of the
world and give speech to the world. This is the special task of the
poet, but ultimately *our* task: "The poet wants the world to concel-
ebrate its praise with him, and he with the world. In this multiple
yes, which the world bears in it and which we bring to song, men
offer the world to each other, make it inhabitable. For a world without poetry is uninhabitable. . . . It is by being sung that the world is
properly a world."[46] As in Zizioulas, we have a creative and poetic
mission, a call to carry the world bodily before God, to make it a
place for dwelling and habitation.

The notion of the call is central to Chrétien's phenomenology and
is closely linked to that of creativity: "What is beautiful is what calls
out by manifesting itself and manifests itself by calling out."[47] This
call calls us out of ourselves and unsettles us.[48] While he articulates

the notion of the call most fully in his *The Call and the Response*, it returns in many of his other works and informs his discussions of corporeality and especially of the vulnerability of the voice. Our speech is always a response to a prior call. Birth itself is already a call. Both call and response are a gift. We truly become ourselves as we respond to the call. The call for Chrétien is in some sense constitutive of what it means to be human.[49] Speech is always a response to a call. Response and responsibility are intimately linked.[50] Like Zizioulas, Chrétien also interprets the call in ecclesial and Eucharistic fashion: He speaks of our care as a "eucharist of speech" and repeatedly calls for us to become the mystical body of Christ.[51] Yet Chrétien makes more positive use of Lévinas by reading the availability of Lévinas's notion of the "me voici" ("here I am") explicitly in terms of responsibility and exposure.[52] Our response makes visible the call of the other. The phenomenology of the call is also linked to the voice, to memory, to the promise, and to hope.[53]

Both Chrétien and Zizioulas, then, speak of a responsibility we have for and toward creation. In both, this is linked to the call and to creativity. And both speak of it in terms of a "lifting up" of creation in "praise." Chrétien puts it as follows: "But if the hymn is vocal, it always has, as well, the scope of an act of witness—the irreplaceable witness of man as such, but also the taking as witness of the entire creation, which itself bears witness in our song. We have to speak on behalf of things, and not only on behalf of one another, as if the world were merely human. Man has a responsibility for creation, a responsibility with which God has entrusted him."[54] Speech becomes an "altar" for the praise of the world. Sometimes Chrétien also veers in the direction of seeing creation as a mere backdrop for human dwelling. By offering the world in praise to God, we render it habitable. Praise fuels our labor in and on behalf of the world, inviting us to care for it. Praise is a response to God's creation. Chrétien suggests in one place that if we were to refuse to praise, the world would be rendered silent and uninhabitable.[55] Praise transforms the world into a good dwelling place for all creatures. Our voice must speak on behalf of creation, as in St. Francis's Canticle of the Creatures: "We can be brothers of the wind and of the moon only through this brotherhood founded in the Word that was itself made flesh . . . it is not enough to sing the world, this song must have meaning, it must say something, it must make sense."[56] We become truly ourselves in this praise on behalf of all.

Yet Chrétien's account is different from that of Zizioulas in some important respects: First of all, the call does not proceed only in one direction and is not given only to humans. In contrast to Zizioulas, for Chrétien God's gift is to the whole world and not just to human beings.[57] Thus although call and creativity are also central for Chrétien and evoke a human responsibility for creation, this is phrased far less in terms of human uniqueness. Other creatures are called as well and can also call us. Indeed, the world responds to its creator: "The world itself is heavy with speech, it calls on speech and on our speech in response, and it calls only by responding itself, already, to the Speech that created it."[58] Chrétien always emphasizes our response to the call. And most of the time we respond not directly to God, but instead respond to the call of beauty that proceeds from the world. "We cannot sing the world," Chrétien stresses, "unless the world itself sings already."[59] The act of praise is precisely an encounter with beauty that calls forth a response.[60] Beauty, also, is not merely a human project for Chrétien.[61] Creativity is not solely about human production of art. Rather, beauty is the very call of the world. We do not make this beauty, but it is always already there. In fact, our "chant" adds nothing to the beauty of the world.[62] The world is filled with a creative beauty that calls out to us. The world and nonhuman creatures are then not mere objects of our care but can themselves become subjects or even real partners in a dialogue. For Chrétien we participate in a larger chorus: The world as a whole offers itself to God.[63] The care and sheltering are mutual on some level in that they can proceed in both directions: from human to creation, from creation to the human being, and together directed to God.

Most importantly, however, for Chrétien there is always an essential vulnerability implied in all human speaking and offering. He stresses that "our voice cannot build the ark of speech, in which everything will be given shelter and received, unless it be in proportion to the hospitality of its silence. It is only by being perpetually translated from silence . . . that our voice can falter and allow itself to be broken, to give itself to what it has to say in such a way that it is not left intact."[64] Indeed, "we offer the world in offering ourselves."[65] It is this brokenness and vulnerability, it seems to me, that preserves Chrétien's thinking of human responsibility from becoming a destructive superiority: in Chrétien our creatureliness is preserved and even stressed. Vulnerability and the need for humility are frequent themes in Chrétien's writings.[66] The voice calls forth a response from the other and offers hospitality to this response in its very flesh: "To see the suffering and beauty of the visible in the form of voice is to be dedicated to providing it forever with the asylum of our own voice. When the eye listens, we must answer what we hear and answer for what we will hear."[67] For Chrétien, this is an act of justice that listens and gives voice to the victim. And it is always marked by insufficiency; the voice is torn and the speech broken.[68] Prayer, as Chrétien insists, is "wounded speech." Prayer gives voice to the suffering of all creatures: "Yes, beauty can say adieu, but it says adieu only when we offer our own adieu, and all of our voice, so that beauty can break it, and make it all the surer for having been broken, and make it remain trembling, with the trembling of the person who knows that all sonorities of the song of the world, without a single one being forgotten or left by the wayside, will have to be given by him, in the eternal dawn, to God, who alone made them arise from silence."[69] Chrétien speaks of beauty itself as effecting a kind of wounding.[70] The difference between Zizioulas and Chrétien is this sensitivity to human creatureliness in all its finitude and vulnerability, but also the recognition that beauty is to be sheltered and protected, not primarily created, by us.[71] Human creativity is a response also to the world and not merely to God. And it is God alone who is the ultimate source of all life and creativity. Human creativity only becomes true care for creation if it does not elevate humanity into a position that absolutely distinguishes it from the rest of creation.

Chrétien does not draw any sort of ecological conclusions from his treatment. Although creation is mentioned frequently in his work and the beauty of the world explored in many places in his writings, he never develops these references into an explicit call for us to take concrete ecological action. Here Zizioulas is far clearer in his repeated injunctions for us to be concerned about the environmental crisis. Such a call must, in fact, give us a sense of our responsibility for the care of creation in order to be convincing as a call to ecological action (or at the very least for refraining from destructive action). Yet, must such a call rely on human uniqueness and distinction from all other creatures?[72] Chrétien's work suggests that human responsibility for creation can be articulated as a response to God's creativity and creation's beauty instead of primarily stressing *our* creativity. As he says, "In order to think responsibility, we must philosophically think what is involved in responding."[73] Responsibility is first of all response, not independent initiative, and it is a response that implies our transformation.[74] The beauty of God's creation calls us to protect and preserve it, to shelter and share it with other creatures. Instead of a new articulation of our uniqueness and distinction from other creatures, be it in terms of our special "call" or as "creativity," this responsibility requires instead a stronger sense of our own finitude, a more poignant realization of our creatureliness, and a humble awareness of our potential for destruction. It is only if human vulnerability and creatureliness are remembered that a human sheltering of other creatures can become genuine care instead of abusive use.[75] We are called to creativity *in service of* our fellow creatures, as we build (and preserve) a habitable world *together*.

Creature Discomforts: Levinas's Interpretation of Creation *Ex Nihilo*

Jeffrey Hanson

The notion of creation *ex nihilo* has understandably been treated as a matter of theological or ontological interest throughout the overwhelming majority of Western history. The last century, though, has seen at least one remarkable effort to rethink creation *ex nihilo* as a component of a meta-ethical philosophy, an effort mounted by no less a thinker than Emmanuel Levinas. Particularly in *Totality and Infinity*, Levinas is not shy about weaving creation *ex nihilo* as a thread into the tapestry of interrelated concepts he wishes to explore and exposit.[1] It is one of the more underappreciated of a multitude of expressions Levinas deploys to refer to the central gesture of his thinking, the encounter with the Other that inaugurates responsible selfhood. The language of creation is surely one of the least familiar and least well-attended-to in the corpus, though, for reasons that are not entirely mysterious. Creation *ex nihilo* is invoked comparatively rarely by Levinas, and the status of such a conspicuously theological trope is potentially confusing in a work that aligns itself with the phenomenological tradition, penned by an author who consistently denied he was a theologian.[2]

The aim of this essay is to give an account of Levinas's argumentative agenda in appealing to creation *ex nihilo* across his published corpus and to raise a final concern about his use of this concept. In the first section I explain the two purposes that Levinas has in mind when he references creation *ex nihilo*—purposes that remain consistent in his work. The second section examines Levinas's fruitful meditation on the family and links the family with his understanding of creation *ex nihilo*. The third and final section studies an essential discontinuity within Levinas's various deployments of the notion of creation *ex nihilo* and argues that while his argumentative agenda when it comes to this issue is consistent across his career, his understanding of subjectivity changes to emphasize the subject's radical passivity. I conclude by raising the possibility that because of this inconsistency the twin points Levinas wishes to make by way of creation *ex nihilo* actually cannot be held together if we follow him all the way through the progress of his career—from *Existence and Existents* to *Totality and Infinity* and on to *Otherwise than Being*—but are best sustained in tandem if we remain with his analysis of the family from *Totality and Infinity*, explained in Section 2.

1. The Double Thematic of Creation *Ex Nihilo*

There are two closely related main purposes Levinas seeks to accomplish by deploying the theme of creation *ex nihilo*. First, it is a way of marking the condition of the ethical subject. The force of the notion of creation *ex nihilo* underscores the absolute character of the ethical demand, responsiveness to which establishes the self in her subjectivity on the basis of an ethical summons that echoes from an unrecoverable past. As there is nothing "before" creation *ex nihilo*, so there is no demand "prior" to that of the Other—certainly not the exigency of being. To be created according to Levinas is to recognize that my true immemorial origin is in responsiveness to the Other, a summons "older," so to speak, than my own preferences and needs. Levinas writes in *Totality and Infinity*, "The unity of spontaneous freedom, working on straight ahead [*oeuvrant droit devant elle*], and critique, where freedom is capable of being called in question and thus preceding itself, is what is termed a creature. The marvel of creation does not only consist in being a creation *ex nihilo*, but in that it results in a being capable of receiving a revelation, learning that it is created, and putting itself in question. The miracle of creation lies in creating a moral being."[3]

Creation is thus not to be understood as a dogmatic doctrine that accounts for the temporal beginning of the world or an ontological theory to demonstrate why there is something rather than nothing,[4] but as an ethical teaching. One is capable of regarding oneself as created, and thus as put into question, at the behest of another. In so doing I do not forsake myself as subject, as Levinas often argues, but I unify critique, the capacity for ethical self-judgment, with spontaneous freedom, the capacity endowed to me by my chronological birth in history, and thus am born as a responsible subject. Creation *ex nihilo* is not an explanation of the subject's past according to Levinas but a reference to its future as determined by a past that can be neither remembered nor completely forgotten.[5]

The second major purpose Levinas has in mind by resorting to creation *ex nihilo* is to reconfirm the ethical subject in her singularity without reabsorbing her into a totality.[6] Causal schemes,[7] analogical relations,[8] emanatory and participatory ontological models[9] are all guilty of such reassimilation of subjects into oppressive conceptual totalities. Levinas declares that "the idea of creation *ex nihilo* expresses a multiplicity not united into a totality; the creature is an existence which indeed does depend on an other, but not as a part that is separated from it. Creation *ex nihilo* breaks with system, posits a being outside of every system, that is, there where its freedom is possible. Creation leaves to the creature a trace of dependence, but it is an unparalleled dependence: the dependent being draws from this exceptional dependence, from this relationship, its very independence, its exteriority to the system."[10]

Only creation establishes a truly separated subject, one that bears the trace of the infinite and a derivative dependence thereon, but in that dependence precisely is certified in its independence. This independence is not validated by freedom or happiness, its capacity for spontaneity. It is a function of the withdrawal of the infinite, which refuses to encompass all in totality. Understood this way, as a statement on the freedom of the subject won by openness to the Other, "the paradox of creation thenceforth loses something of its audacity."[11]

The paradox of creation appears less audacious when we revise two commonly accepted notions of the infinite and the finite respectively. First, the infinite is often assumed in its limitlessness to be an ally of or a figure for totality. To be infinite is to be all-encompassing, and the finite conversely is allowed only to be a moment within the infinite. Second, the finite is often thought of as a mere "diminution" of the infinite, a "fall" from or "degradation" of the perfection of the infinite. According to Levinas, both of these commonplaces are to be rejected. Absolutely key to understanding this twofold rejection is the recognition that it follows from Levinas's decisive repudiation of the procession of essence from the One "by emanation and by *descent*," a trope he associates with both Plotinus and Parmenides.[12] Instead, Levinas follows Plato, who he argues "posits transcendence as surpassing the totality"—that is to say, he reverses the logic of emanation that both preceded him and resurfaced after him.[13] The infinite according to Levinas himself and his reading of Plato admits "a being outside itself which it does not encompass" and indeed accomplishes its very infinitude precisely in remaining proximate with a being that is itself positively separated from the infinite.[14] The infinite therefore must be thought of as accommodating itself to the separated being; in fact, this accommodation is the very means of the infinition of the infinite.[15] As Levinas writes, "Infinity is produced by abandoning the invasion of a totality, in a contraction that leaves a place for the separated being. Thus relationships that open up a way outside of being take form. An infinity that does not close in upon itself in a circle but withdraws from the ontological extension so as to leave a place for a separated being exists divinely. Over and beyond the totality it inaugurates a society."[16]

Notice the emphasis that Levinas places upon the relations between infinite and finite being "over and beyond" the totality. The point is that the infinite does not deplete itself in accommodating itself to the separated subject but remains "beyond" in its very accommodation. Likewise, and this is the second reinterpretation gestured toward earlier, the separated subject is realized positively and sufficiently unto itself—though to abstract thought, which focuses on the finitude of the subject, it will seem that "the relations with the Good are announced through an apparent diminution."[17] As he affirms throughout *Totality and Infinity*, so Levinas repeats here his insight that "the relation connects not terms that complete one another and consequently are reciprocally lacking to one another, but terms that suffice to themselves. This relation is Desire, the life of beings that have arrived at self-possession. . . . Separation is the very constitution of thought and interiority, that is, a relationship within independence."[18] The language of relationship within independence of course recalls the apparent paradox of independence within dependence, and the reference to desire recapitulates one of the major themes of the text as a whole.

The key theme of desire is explicitly connected to creation more than once by Levinas: "But the order of Desire, the relationship between strangers who are not wanting to one another—desire in its positivity—is affirmed across the idea of creation *ex nihilo*. Then the plane of needy being, avid for its complements, vanishes, and the possibility of a sabbatical existence, where existence suspends the necessities of existence, is inaugurated. For an existent is an existent only in the measure that it is free, that is, outside of any system, which implies dependence."[19] Here again creation is linked overtly with desire in its positivity, not as lack but as the solicitation of the infinite of a being already sufficient to itself, drawing satisfaction from and within its very needs, and the exteriority to system and totality that creation necessarily implies.

2. Creation and the Family

To complete the picture of creation we see rendered for us in *Totality and Infinity*, it is necessary to say a few words about paternity, which is also plainly related to the theme of creation. Paternity is a way of naming the fruition of desire, the issuance of multiplicity without totality, and unparalleled independence in dependence, inasmuch as the son does not merely issue from the father but emerges as other to the father. As Levinas writes early in *Totality and Infinity*, anticipating his concluding remarks on the subject of paternity: "The great force of the idea of creation such as it was contributed by monotheism is that this creation is *ex nihilo*—not because this represents a work more miraculous than the demiurgic informing of matter, but because the separated and created being is thereby not simply issued forth from the father, but is absolutely other than him. Filiality itself can not appear as essential to the destiny of the I unless man retains this memory of the creation *ex nihilo*, without which the son is not a true other."[20]

That creation *ex nihilo* is not a matter of the son simply being "issued forth from the father" is consistent with Levinas's rejection of creation by emanation, or the self-unfolding of totality into a merely apparent otherness or an otherness that is alienated from itself, only to be dialectically recouped in the end. It is likewise consistent with his insistence that the infinite does not impose itself upon the finite but retracts from the finite, which itself is not a diminution of the infinite but is a positively constituted separation, independent in its dependence,[21] in no way relative to the infinite.[22] This dynamic is discernibly at work in the relationship of parent to child, though as we shall see there are arguments that could be made in favor of both the child and the parent being created; both could in a sense be said to partake in creation as absolute relation between same and other.

The leading exponent of Levinas on creation is without doubt Michael Fagenblat, and in his book *A Covenant of Creatures: Levinas's Philosophy of Judaism*, principally on the basis of the analysis of fecundity, he reads the subject of *Totality and Infinity* as a "moral creator" who gives way to "the ethical creature" of *Otherwise than Being*, along with a great many other dramatic conceptual shifts that in his reading sharply distinguish the two works.[23] Speaking to this perceived shift, Fagenblat writes, "Unlike the moral subject, the ethical creature is exposed to the other prior to its own self-identification. The other lives 'in me and in the midst of my very identification,'[24] before the subject can assume its moral commitments and accomplish a moral world, in short, before creation. The passivity of responsibility renders the ethical self a creature rather than a creator. As Levinas says, 'the oneself is a creature, but an orphan by birth or an atheist no doubt ignorant of its Creator, for if it knew it it would again be taking up its commencement.'[25] The metaphysics of creation is now regarded as a fantasy of power, separation, and autonomy. To become an ethical creature is to dispel oneself of the fantasy of being a moral creator responsible for one's free acts. The metaphysics of creation—ex nihilo and ex hylus—is thus cast aside."[26]

Fagenblat's sharp distinction between the two great works is clearly delineated but also overstated, as is his concluding claim that the metaphysics of creation is cast aside in *Otherwise than Being*. As we will see shortly, there are references in the latter work to creation, fewer to be sure than in *Totality and Infinity*, but present nonetheless. The contention of the concluding section of this essay is that for Levinas the notion of creation *ex nihilo* throughout the authorship signals the twin themes identified earlier—the condition of the moral subject and the nontotalizing unity of multiplicity— and that the only substantial change he makes to the way in which creation is presented is with respect to the degree of passivity that he ascribes to the moral subject. The moral subject is always a creature (though in fecundity he may imitate the creator in an important way), and what changes over the course of Levinas's thought is how passive the creature must be in order to be considered a creature. This change is an intensification in the passivity of the creature: in *Existence and Existents* the creature is construed as contributing in some way to her own creation; in *Totality and Infinity* there may be some ambiguity about the degree to which the creature can be called "active"; while in *Otherwise than Being* all doubt is removed and the creature is called more passive than passive. Fabgenblat is right that in the latter text metaphysics is cast aside (rather than rehabilitated as in the earlier text), but nevertheless in *Otherwise than Being* it is creation *ex nihilo* to which Levinas turns in order to furnish himself with an alternative to metaphysical speculation.[27]

The account of fecundity in *Totality and Infinity* suggests that in childbearing there is a way in which a person can be like the creator. Fagenblat is convinced of the importance of the theme of the family for *Totality and Infinity*: "The family is a model for how love and the desire for goodness are oriented toward particular persons. The fecundity of ethical time is thus the key to the teleological structure of *Totality and Infinity*. This fecundity attests to the desire for the particularity of the other that goes further than the will to knowledge, which abstracts from the singular, goes beyond the quest for authenticity, which forecloses the significance of a future belonging to someone else, and rises above the anonymous values of the State."[28] This is certainly true, and language such as Fagenblat's here recalls the importance of the family in connection with the second theme to which creation *ex nihilo* is directed in Levinas's thinking: toward articulating a community of singulars without reducing them to a totality. As he says on the same page, "In envisaging an eschatological peace as the purpose of creation, what is important for Levinas is to sustain the singular, embodied, and unpredictable character of the other whose horizons exceed one's own sense of the possible. He therefore boldly, if problematically, asserts that the ethical desire for the particularity of the other is 'concretized in the marvel of the family.'"[29]

What this reading suggests is that fecundity and paternity are models of creation insofar as the child is a genuine Other; the child realizes the purpose of creation in her singularity, her embodied and unpredictable character, in her capacity to surprise (her parents especially, one presumes), and her survival beyond her parents' death.[30] The family is the concretization of the two key themes embraced by the notion of creation. It is the drama, in Levinas's special sense (following Nietzsche) of that term,[31] of creation realizing its purposes—the moralization of the singulars and the establishment of their shared nontotalized multiplicity. The linkage here is more than speculative. At least twice the connection between family and creation is asserted directly, once in the quote earlier from page 63 of *Totality and Infinity* and once more in the section on "Filiality and Fraternity:" "Creation contradicts the freedom of the creature only when creation is confused with causality. Whereas creation as a relation of transcendence, of union and fecundity, conditions the positing of a unique being, and his ipseity qua elected."[32] In this passage creation is explicitly called a relation of transcendence (as opposed to the relations analogous to those of transcendence described in Section II of the same book),[33] and at the beginning of the text Levinas referred to the final meditations from which this passage is excerpted as "conjunctures in being."[34]

Section IV is a clear, almost point-for-point, in fact, reflection and correction of the dynamics of enjoyment explored in Section II.[35] As an example, consider the way in which the child being "mine" is opposed to the way in which the products of my labor are "mine." I can refer to my son as "mine," but this mineness is completely different from the manner in which the products of my labor are mine, a mineness that is both more proprietary and more provisional. "The son is not only [*pas seulement*] my work, like a poem or an object, nor is he my property. Neither the categories of power nor those of knowledge describe my relation with the child."[36] And again, "He is not only [*pas seulement*] my work, my creature, even if like Pygmalion I should see my work restored to life."[37] By contrast, the fruits of my labor are mine on the condition that they lose their in-itself character and become immediately exchangeable for arbitrary monetary value, something that could never be said of a child.[38] While labor thus remains within the orbit of the Same for Levinas,[39] paternity opens toward the infinite future of a world that perdures beyond and without me, after my death, and issues ultimately in that which is "beyond the face," "the engendering [of] good being."[40] The good being borne by the father in the son is a true other as well as a figure of creation *ex nihilo*.

So there may be a sense in which the parent is the creator of the child, a promoter of the aims of creation through the fulfillment of the command to be fruitful and multiply.[41] Reading more deeply, though, we can discern a sense in which the *parent*, at least as much as the child, is created *ex nihilo* precisely by the act of bearing and rearing children. There does seem to be something strikingly counter-intuitive about the claim that the child is created *ex nihilo*, since it is rather commonsensical to suppose that she has been created from the act and material of her parents. If we keep in mind, though, that creation *ex nihilo* is first a matter of the moral condition of the ethical subject, not an explanation for the ontological origin of anything, it becomes much more intuitive to suggest that the egoism of the parent (even the egoism of two that can obtain between erotic lovers) is decisively broken by the advent of the child.[42]

Without endorsing a position as extreme as that the child is *in no sense* a creature, we could argue that the parent is *in a sense* a creature, in that the parents' being is radically altered by the drama of fecundity. This alteration is accomplished in a gesture that Levinas does not hesitate to dub a "trans-substantiation" effectuated by "amorous subjectivity"[43]: "In this unparalleled conjuncture of identification," he writes, "in this *trans-substantiation*, the same and the other are not united but precisely . . . engender the child."[44] Trans-substantiation here serves as an equivalent term to being created *ex nihilo*, though quite clearly it is the parent who is now good, responsive to the Other that is the child, and thus for itself. Repeating that the reality of creation is a relation, Levinas writes, "In the conjuncture of creation the I is for itself, without being *causa sui*."[45] The parent is for itself and thus good and at the same time is not *causa sui* but fundamentally in relation.[46] This again is a reminder that creation *ex nihilo* for Levinas establishes the ethical subject in both its integral unity and relationality without totality; in this exemplary instance, the family is the nontotalizing community of singularities, and the family involves not just the children but the parents as creations, as well.[47]

3. The Destiny of Creation

This final section of the essay examines the curious destiny of creation *ex nihilo* as Levinas reworks his theme from the beginnings of his career in *Existence and Existents*, where creation *ex nihilo* is rejected as too passive to capture the sense of subjectivity Levinas is exploring in this early work up to the last of his great works, *Otherwise than Being*, where creation *ex nihilo* is singled out as the only idea that captures the radical passivity of the subject as Levinas comes to present it in its final form. In closing, I identify a potential problem for Levinas: The more Levinas attributes passivity to the subject, the more he lets slip a satisfying account of the togetherness of ethical subjects. At the end of this essay I suggest a return to the drama of the family, exposited earlier, as a more helpful way of preserving both of the two roles that creation *ex nihilo* plays in Levinas's thought, though Levinas himself seems to abandon the family in his final major work in favor of an account of justice.

To fully appreciate what Levinas means by creation *ex nihilo* from the beginning of his career to the end, we must take stock of what is meant by "nothing" in *Existence and Existents*, a study that will shed the final light on this subject in Levinas's writings. In *Existence and Existents* we read the famous thought experiment in which Levinas invites his reader to imagine the reduction of all things to nothing; relying on Heidegger's articulation of the ontological difference, Levinas claims that the reduction of all beings, all existents, to being or existence yields not nothing as nonbeing but precisely nothing as sheer, undifferentiated being, the "there is [*il y a*]." The "there is" is the impersonal, featureless, and undifferentiated fabric of being, and its vacuity is revealed precisely as an intimation of nothingness, where nothingness is only a "profile"[48] of being itself, such that "the fear of nothingness is but the measure of our involvement in Being,"[49] and conversely, "being insinuates itself even in nothingness."[50] So when Levinas speaks of creation *ex nihilo* it is probable he does not have in mind the absence of being but a suffocating too-much-ness of being itself that is revealed as no-thing, the literally inarticulate horror of Being's inescapable oppressiveness.

This too reinforces the primarily ethical meaning of creation *ex nihilo*; to be created "from nothing" is to be extracted from being, to be absolved of the tyranny of essence. Interestingly, Levinas makes just one passing reference in this text to creation *ex nihilo*, and here he discounts it on the grounds that it implies a "pure passivity on the part of the creature."[51] At this early stage it would seem Levinas attributes a share to the creature in her own creation—via the hypostatic upsurge that is the event of subjectivity; according to the hypostatic dynamic I insert myself within and against the "there is." Much later in his career, when in *Otherwise than Being* he reinvokes creation *ex nihilo*, he does so because it is, he claims, the only figure of thought adequate to the task of accounting for creature status as an "absolute"[52] and "anarchical"[53] passivity "more passive still than the passivity of matter."[54] It is noteworthy then that in the course of his writings, Levinas moves from regarding creation *ex nihilo* as insufficiently radical to the task at hand in *Existence and Existents* of accounting for how a singular emerges from the anonymity of being, to regarding it as the only gesture radical enough to the task at hand in *Otherwise than Being* of demonstrating how the singular draws its meaning from a source more ancient than being.[55]

One potential area of concern for Levinas's account is that if we take this final late reference to creation *ex nihilo* in *Otherwise than Being* seriously, then what we discover is that the more he reinforces, as he does, the first purpose of the theme—to establish the meaning of the ethical singularity—the more he lets slip a robust defense of the second purpose of the theme—to describe how ethical singularities form a nontotalizing multiplicity. More precisely, the more Levinas insists that the ethical subject has nothing of her own apart from and prior to the bestowal of her subjectivity upon her by the Other, the more he requires but does not in fact provide something like the family, a community of singulars that stand in meaningful ethical relationship to one another without totalization.

The recourse to creation *ex nihilo* occurs at the outset of the crucial chapter entitled "Substitution," and it does so in direct response to Levinas's rhetorical question as to whether the analysis heretofore has been sufficiently faithful to the "anarchy of passivity."[56] "It is perhaps here," he writes, "in this reference to a depth of anarchical passivity, that the thought that names creation differs from ontological thought . . . in creation, what is called to being answers to a call that could not have reached it since, brought out of nothingness, it obeyed before hearing the order."[57] Consistent with the whole thrust of *Otherwise than Being* in its divergence from *Totality and Infinity*, the emphasis is here not so much on the unity of spontaneous freedom and self-critique but on a voluntariness prior to "the will's initiative (prior to the origin)"[58] and an already accomplished exposure to the demand of the Other. Notice now that the subject is called "to being," not out of nothing as a profile of being, and she answers a call she has not yet even heard.

Whereas in *Existence and Existents* there was a clearer role for the subject to play in her own creation, via the hypostasis or event of involution within the seamless fabric of being that establishes the separated, atheistic self, now it is clear that creation *ex nihilo* does not extract the ethical self from being but disposes her "beyond essence,"[59] even in such a way that this disposition much more emphatically than in *Totality and Infinity* "precedes essence in" the subject herself and thereby accords the self no role in creation.[60] In the earlier text Levinas insisted that an account of the creature must include an explanation of "what in it receives existence";[61] in the later text "creature status" consists in the "'hither side,' [*en deçà*] which a being retains no trace of."[62] The work of creation is done entirely by the other-in-oneself, according to the demands of the "most passive passivity."[63]

For reasons that are familiar to readers of Levinas, he chooses in *Otherwise than Being* to reorient his argumentation about the nature of the ethical subject in a direction that diminishes the impression he left in *Totality and Infinity* that the subject has any privilege, even a curtailed one, apart from the inhabitation of the ethical subject by the Other. The atheist subject does not await an encounter with the Other in an undetermined future; the Other is already within the subject from an immemorial past. This overall transformation in his thinking affects Levinas's specific deployment of the trope of creation *ex nihilo*, as well. Gone are the references to nothing as profile of being, gone are the implications that separation is a necessary precondition of the encounter with the Other; instead, the impression that the subject has nothing—literally nothing, no purchase within the field of essence—of her own apart from what is invested to her by the Other is hyperbolically intensified.

And this shift could very well be defensible as a strategy to secure the first purpose of the use of creation *ex nihilo*; it could be seen as a necessary clarification of the exact status of the ethical subject, a status that in *Otherwise than Being* dispels the ambiguities identified in *Totality and Infinity* by Jacques Derrida and others.[64] As a consequence of this strategy, it is necessary to point out that creation *ex nihilo* now must be understood in a different way than it was presented in the earlier work, a way that in keeping with the revised architectonic of the book as a whole deprivileges the role of the subject in the process of creation. If that is so, then the consequence Levinas has to face is that it becomes more difficult to use creation *ex nihilo* as a device for securing ethical singularity without something like a more developed scheme for uniting a community of singulars in their mutual interaction with each other; the more consistently he pursues the first purpose of having invoked creation *ex nihilo*, the less clear becomes the means by which the second purpose is realized. In a more traditional way of thinking, the doctrine of creation *ex nihilo*, understood as the teaching that the creature has nothing of her own she has not received from the creator, necessarily entails that she thereby shares in the gift of being with all other creatures. This is a conclusion that Levinas cannot support in its classical formulation, because of his opposition to all forms of analogy and participation.

Instead, the answer that Levinas provides in *Otherwise than Being* to the question of what binds singulars together as independent creatures in a nevertheless nontotalizing manner is justice. This account has struck some readers as unsatisfying, and one reason for misgivings is that it reinvokes the apparatus of totality—the state, politics, techniques[65]—and attempts to reinscribe them with the Other in the Same in a way that is arguably not fully worked out. In a highly problematic fashion Levinas at the end of the text reasserts the requirement for justice as explicitly a "question of consciousness" and restores the dignity of "intentionality," "visibility," "system," and "thematization" as purported functions of the face-to-face.[66] Suddenly "the asymmetry of proximity in which the face is looked at" is subjected to "incessant correction" by the third party, and "my anarchic relation" undergoes a "betrayal."[67] In the end we find that "the comparison of incomparables" is called for, such that "the order of justice" is now "moderating or measuring the substitution of me for the other,"[68] without Levinas ever even hinting that such an operation will produce righteous outcomes. In short, Levinas invokes justice but cannot guarantee that justice will be just.[69]

The dependence of *Otherwise than Being* on justice does not just fail to capture the second purpose creation *ex nihilo* was meant to serve but in fact runs athwart it. As an alternative, a return to the analysis of domesticity and fecundity from *Totality and Infinity* seems more promising. More readily than justice, the family provides a nontotalizing way of respecting the singularity of all singulars while accounting for their togetherness.[70] For the Levinas of *Totality and Infinity*, the meditations on fecundity and domesticity promise the possibility of keeping his twin objectives preserved together more readily than the appeal to justice in *Otherwise than Being*. It is through the bearing of children that the truth of the beyond essence is perpetuated, and it is in the home that the welcoming of strangers is concretely enacted. It is no accident that these meditations are in no way recapitulated in *Otherwise than Being*, nor is it an accident that hospitality is replaced by the experience of being "hunted down even in one's home,"[71] nor that this text specifically denies that ethical subjectivity is a matter of transubstantiation.[72] A reappraisal of the family might recover a broadened sense of what Levinas was pleased to call in *Totality and Infinity* "good being." Creation *ex nihilo* implies that there is nothing essential about essence. The more rigorously we grasp this truth, the more we will be able to echo the founding words of creation: "It is good."

Reflections from Thoreau's Concord

Edward F. Mooney

> To be a philosopher is not merely to have subtle thoughts . . . but so
> to love wisdom as to live according to its dictates, a life of simplic-
> ity, independence, magnanimity, and trust. It is to solve the prob-
> lems of life, not only theoretically, but practically.
> —*Henry David Thoreau*, Walden

With Thoreau in mind, I set out to reflect on creation, on being a
creature among others in creation, and on the role of creativity as a
vector animating creatures and creation. These might be philosophical or theological matters, but in any case, Thoreau would have me
look beyond my would-be subtle thoughts on such august things.
He'd have me ponder smoothly or subtly or awkwardly within the
venture of my *living as such*—the life I'd *call* mine (and *would* be
mine) here and now. A philosopher will *live into and live out and
live from* such reflections—as if living were thinking-imbued or
thinking-saturated—and as if thinking without living were blind,
monstrous, or misshapen. He doesn't forbid "subtle thoughts." He
had quite a number to convey, especially on creation and creatures.
But subtle thought is not yet philosophy, not yet love of wisdom,
and he was interested in what it takes to be wise or a philosopher,
what it is to live out of, or into, or alongside these thoughts.

In a surprising and swift stroke, Thoreau defines the love of a
would-be philosopher. We might be jolted to discover that in his

view, wisdom and philosophy are *not* simply love of knowledge for its own sake. Thoreau is specific. If we seek wisdom, and philosophers must seek wisdom, then we seek *a way of life.* Thoreau names the strengths that a philosophical life will exhibit. As philosophers, we will yearn to live in *"simplicity, independence, magnanimity, and trust."* Such mindful, alert philosophical living—in his essay "Walking," he calls it "sympathy with intelligence"—will be living infused with creativity. For Thoreau the task is to live as one among innumerable creatures in creation, in Nature, sympathetically and alert, with simplicity, independence, magnanimity, and trust.

This way of living does not have to be esoteric or heroic. Of course, we inhabit such virtues as Thoreau recommends always in part, and always with beckoning of better and more vibrant ways to integrate them into a life. But still, they can become present in small and telling ways.

I remember a moment at a continental philosophy and theology meeting at Gordon College, not far from where Thoreau's Concord River meets the Merrimack and the Merrimack meets the sea. A few of us left half-downed beers and headed for second- and third-floor rooms. There was a slight pause at the elevator, as if a collective decision were impending. Then in an instant, all turned for the stairs and bounded up. That was *independence* (this bounding was not herd instinct). It had *simplicity.* And in phases of life other than bounding up stairs, I doubt not at all that my friends are *trusting* and *magnanimous.*

Conversation

As we read *A Week on the Concord* or the *Journals* we are addressed. Words arrive strewn like seeds in the wind, finding multiple fields to take root. In some instances they bloom over the years and around the world. Think of the century-long and global flowering of his essay on Civil Resistance. We might also imagine that we are addressed here and now by Thoreau. Imagine what he would think or say to address us directly, were he to walk in through the door in

the rear of this hall in Southern California, bringing him present by
a different route, through the address of his words. If he entered the
room, I imagine him searching for words, wondering what to say—if
anything. I suspect he would peek in, realize he had taken a wrong
turn, and head back to the beaches near Carmel or up to a tent on
Mt. Baldy. He is curious, attentive, all ears and eyes, but academic
discussion is not his cup of tea. And he might be afraid we'd corner him with *sensationally bad news*—with tales of species extinction and other eco-catastrophes. How much of the spectacular, how
much bad news, how much of the banal, can he stomach?

Thoreau lamented the shrinking of the salmon population in
Concord River, but he would not have traveled to Boston to lobby
against dams or for limits on the catch. Not that he was a political
quietist. He just wasn't a lobbyist or campaigner. Politics should not
be all-absorbing, he thought. That didn't mean he'd forgo working
the Underground Railroad, speaking out in political crisis at rallies,
spending a night in jail, or ringing the bell for John Brown. But a
daily barrage of negative news saps life.

Our Place: The Concord

The Musketaquid, or Grass-ground River, though probably as
old as the Nile or Euphrates, did not begin to have a place in
civilized history until the fame of its grassy meadows and its
fish attracted settlers out of England in 1635, when it received
the other but kindred name of CONCORD from the first planta-
tion on its banks, which appears to have been commenced in
a spirit of peace and harmony. It will be Grass-ground River
as long as grass grows and water runs here; it will be Concord
River only while men lead peaceable lives on its banks.

Thoreau, A Week on the Concord and Merrimack Rivers

We know where to find Thoreau, on the Concord, or at Walden,
or on the way to Maine's highest peak, Ktaadn. But where, exactly,
is *our* abode? We do not quite find ourselves on the banks of the

Grass-ground River, or on the slopes of Mt. Greylock or Ktaadn. Yet we seem to follow Thoreau's life on the way to and among these sites. Being with him through words, do we become residents-at-one-remove? Let's say that answering his address, traversing the landscapes he traverses, is a way to abide with him, reside with him. If we accept this picture of our contemporaneity with him, we may discover something remarkable about his identity and about our identity as we abide with him.

We rather automatically think of who we are in terms of something inward—our thoughts, or memories, our private "sense-of-self," our dreams and aspirations, our hidden shames. But Thoreau has a rather different emphasis. He would agree with Emerson, who announces with brief and startlingly finality, "He is place." Granted, Emerson may say other things that seem to subvert the finality of this report. Emerson's thought, like Thoreau's, can lean in multiple directions. They are "experimental," improvisatory thinkers who experiment at the extremities of thought. They abjure fashioning building-block thoughts finished off for assembly into systematic, orderly, and quite stationary wholes.

Let me return to the strangeness of Emerson announcing that "he is place," and to the wider issue of how this suggestion, if true, helps us understand how we travel with Thoreau if we travel with the places he passes on to us. What startles us if we entertain the thought that *we are place* is that with that move, everything "inner" or "private" or "hidden" seems to become irrelevant.

Emerson does not say that I am a soul or spirit or a set of thoughts and aspirations neatly located *in a* place, or that to flourish I must find *a* place of flourishing. If he had, he would have continued an exterior/interior contrast and merely asked us to remember that these terms form a communion: I (my interior) reach fullest expression in the landscape of my activities (my exterior). But Emerson (and if I am right, Thoreau) are letting us entertain a more radical proposition. We are always "exterior," which radically transforms our sense of identity and our sense that the "exterior/interior" contrast is *the* inescapable bottom-line contrast for thinking about identity.

Emerson says that as persons we *are* place: no more, no less.[1] Can we get an intuitive sense of this? Perhaps the idea is that I am the room that receives my voice, and the valley that receives my gaze; I am the cup that receives my spirit, and the arm casting up toward the sky and falling down to the depths of the pond. I am the landscape that welcomes my eyes. If alive, I am place. The matter of place is a matter of identity. In fact, the "I am" seems to melt away. Place is all-present: room, window, rapt listeners, smell of caffeine, rustle of paper, podium. The perceiving "I" becomes as redundant and superfluous as the nametag attached to my lapel: place is all.

It's easiest to think of Thoreau as *a* creature *in* creation—and no eyebrows will rise at that suggestion. But let's say he *is* creation. That might seem to be making him too cosmically important, and if he were to think it, make him something of a megalomaniac. But perhaps it's to say that in reading him, what we retain is the allure of the place, the allure of a meditation of the place, where the rustle of sky or the darting of fish let "the I" of the writer become as redundant and superfluous as the nametag on the cover of the book: a filing device, but no substitute for the wondrous places that become present page after page.

To entertain the thought that place is all is not to say that Thoreau is temporally and spatially coextensive with all that has ever existed, and in that sense "*is creation.*" That would be megalomania indeed! The thought that Thoreau is place, and by extension, that he is creation, might be instead a gesture more of humility than grandiosity. First, Thoreau is one among trillions with equal claim to utter, "I am creation, I am the swirl of life." Each can become lost in an infinite surround. This thought puts him on a par with each grain of sand washed by the tides. If anything, this thought becomes a *deflation* of any self-standing, self-reliant importance. In entertaining the thought that he is creation, Thoreau would not be claiming to *rule* creation, or to become *equivalent* to the aggregate, blooming-buzzing carnival and funeral of it all.

Second, "I am creation" is humbling (rather than megalomaniacal) because Thoreau absorbs the sobering and difficult fact that he *won't stay forever.* As Thoreau sees it, creation is shot through with ongoing demise and death. Even in the most vivid moment of creation and its creatures is but a passing affair. Creativity is riding this upsurge into life with delight and celebration—and rides the downfall without melancholy or despair.

Continental

Thoreau keeps the company of Goethe and of the *Gita*, of Diogenes, Epicurus, Milton, and Hamlet, as well. That makes him non-analytic, but not quite continental. Pierre Hadot writes an essay on Thoreau and teaches Foucault that philosophy is a spiritual exercise, a way of life. Could Derrida cash out his subtle texts as a pilgrimage walk with Thoreau or as an echo of Diogenes in telling Alexander to get out of his light? *Writing and walking* were spiritual exercises for Thoreau. His writing won't be mistaken for run-of-the-mill continental. He wrote clarion sentences as clear as chanticleer's cry.

I don't want to paint with too broad a brush. Some continentals produce jargon-free sentences as clear as a bell, and some let their writing mix in with their non-scholarly lives, letting subtle thought and subtle living merge, as they should. Our colleague Norman Wirzba writes rural ethics and plows meadows, and our friend Brian Treanor writes on Thoreauvian virtues and climbs rocks in wild country. I'm sure we have a long-distance bicyclist and foot-racer here too, and many who write clear as a bell.[2] We're not just continental theory hounds or Melville's scholarly "sub-sub librarians." We bound up stairs and open our classrooms to salmon, rocks, and turtledoves.

Thoreau liked libraries but not universities. Harvard did not impress. Independence, magnanimity, simplicity, and trust were not on the campus map. In any case, Thoreau did not think that wisdom is love of knowledge. He made full use of the Harvard library and absorbed plenty of what we could call academic knowledge—the histories of peoples, the classifications of plants, the natural history of Cape Cod. But having knowledge was not enough to have wisdom.

Wisdom demanded a kind of living above and beyond one's accumulated knowledge. And insofar as continental philosophy is so often
in a mode of suspicion or critical of modernism, onto-theology, enlightenment, and the like, Thoreau doesn't seem to fit the bill.

Thoreau pegged wisdom as living from and to independence, magnanimity, simplicity, and trust. It's not clear that this tilt to his philosophy helps him to be either continental or noncontinental, An-
glophonic. If both contemporary "styles" maintain a commitment
to knowledge (as critique, as insulated from living), following up a
twist in his sense of living, his philosophy creates further complexity in labeling him for contemporary purposes. Consider his essay
"Walking," where adds another factor that separates philosophy-as-
knowledge from philosophy as wise living.

The list that moves from independence to trust is a record of clear
and present virtues necessary to wise living. What "Walking" adds
to the picture is not exactly a virtue but a broader and encompassing aspirational craving. By all accounts, we are perceiving animals,
and by many accounts, we are knowing animals. Better perception
leads to better knowledge, and our distinguishing mark as creatures
is our better knowledge. So what is more natural than the assumption that the wise person aspires to better knowledge? But in "Walking," Thoreau puts attentive perception in the service of something
other than knowledge. He puts it bluntly: we perceive not to know
but to attain *"Sympathy with Intelligence."*[3] Perception is in the
service of sympathetic immersion, and such perceptual sympathetic
immersion is guided by intelligence. Intelligence focus our region
of attention, and in so focusing our attention sympathetically, the
object of our attention speaks back its intelligence. It's as if to get
at the reality we seek, we must have sympathy and be receptive to
whatever reality, the *intelligibility* of reality, can relay our way.

Sympathy with Intelligence

Why does Thoreau follow the habits of birds and muskrats and date
the honeysuckle bloom? Intelligent observation and scholarship is a

step toward *sympathy* with wonders of creation, an immersion that transcends prior counting and theorizing. And sympathy precedes our counting by directing us toward the idea that it might *matter* to count. Wise living is intelligent, independent, simple, magnanimous, trusting—with sympathy for *this* particular and its surrounding *All*. Thoreau reads to get dispassionate distance on things yet also as a release from the lifeless, the bland and beneath notice. Sympathy brings the dead back to life, the neglected back to center stage—the inanity of noise dies of neglect.

How would Thoreau know, if he were here, that the swamp is a wondrous place and the mall isn't? He would show you, lead you there, let you develop a nose for it. Through sympathetic immersion you get a sense of the swamp's foot-slosh and of what an escape from *escalator precision* is like. To sense a swamp's wonder is being intelligently alert there, finding sympathy for it and its creatures in a way that repays attention as the place brings *you* alive.

The mall, in contrast, deadens the "you" of the wild. It relentlessly takes the measure of your pocketbook, your looks, your taste in accessorizing. In the formerly *dismal*—now *wondrous*—swamp, you resonate with things that breathe, that eat, that live and die, that procreate, that are wet and smell and move and swim and stand and sway. It is replete with living so close to our own that sympathy is indeed rich and rousing.

Such is the circle of the wild we inhabit. The life of the wild rouses the wild life in us. Free of politics, free of metallic or plastic clank or snap, full of raw earthy life. Nothing is priced, everything priceless. The stainless cube of the mall can't yield these resonances, can't yield this vivid palette of sympathies, where our body comes alive with the stream, the leaf, and the feather.

Thoreau calls this sympathetic understanding, or Sympathy with Intelligence. We might call this a "higher knowledge" and place it as the third rung on a ladder: knowing *that* a bare fact is true, knowing *how* to act or cope, and knowing *viscerally, passionately*, where to know presupposes immersion, feeling, and tactile intimacy. This is a knowing that gets degraded and dismissed with the onset of Galilean mechanical science and a Cartesian dualism that strips objects of life. It is a biblical, carnal, poetic, and saving knowing. We know *that* rain is essential to crops, and we can also know *how* to plow and to harvest. And we know *in our bones* as we walk the rhythm and feel of soil, sunlight, and rain—and in that visceral knowledge, feel exuberant kinship with fellow tillers, reapers, creatures, flowing in currents of life and death.

Love and Loss

The phrase *"Sympathy—with intelligence"* reminds me of Nietzsche's advice, that in seeking meaning it is not enough to be "dynamite"—to blow up the false and illusory, to set out on the paths of interminable critique.[4] We should, he tells us, assemble those things around us *that we love.*

> What have you truly loved up to now, what has elevated your soul, what has mastered it and at the same time delighted it? Place these venerated objects before you in a row, and perhaps they will yield for you, through their nature and their sequence, a law, the fundamental law of your true self.[5]

On his walks Thoreau sought those things he could love, even those things that had slipped away and more or less died to him—the fish under ice, the birds not yet returned from the south, the blossoms not quite ready to bud forth again. He searched the things there and the things absent, the fox cavorting across the ice and, more metaphysically, the lost hound and bay horse. "I long ago lost a hound, a bay horse, and a turtle dove."

The eclipse of God is always partial, passing with each season, and never permanent. The dove is lost, but still heard, and chanticleer will crow. We need this unfashionable love of the world— the faith that eclipse is eclipse, not permanent dark, that neither joy nor serenity has met its final hour. Thoreau knew species extinction and would hate farmed fish, but vowed at all costs to reject—*refuse*—dejection.

He had little hope for the future—who *could* on the eve of that Great Uncivil Slaughter? Yet he refuses, he says in *Walden*, to write an Ode to Dejection. He wanted his eye disciplined, vigilant, on the scene before him, in a kind of Stoic awareness in which "nothing is lost." He had lived through the catastrophe of his brother's violent death, of "little Waldo" a month later, of John Brown some years later. He lived through the humiliation of the black tradesman Anthony Burns, recaptured in Boston pursuant to the Fugitive Slave Law. He didn't *preach* serenity and joy. He found them under leaves and in a heron's flight—in the least meadow or bobbing cranberry in the marsh. His brother John died an agonizing death, yet was also serene, grateful to have lived.

> gratitude for life having been given at all is the spring of remembrance, for a life is cherished even in misery. . . . What ultimately stills the fear of death is not hope or desire, but remembrance and gratitude.[6]

These are the words of a young Hannah Arendt, writing on *Love and Saint Augustine*. They might have been Thoreau's.

Without too much of a stretch, one might to see the sentiment Arendt proposes as linking philosophy to gratitude, to remembrance, and to cherishing life, and see this as suggestive of philosophy, living well, as a kind of know-how—knowing how to live with serenity and gratitude. If we link living well, or wisdom, to "know how," then in seeking wisdom we would not focus on cognitive propositions—we could maintain a sort of Socratic ignorance here. Nevertheless, we would devote full attention to living skills that, in this case, turn out to be a weave of exuberance, gratitude, and tranquility.

For what it's worth, Thoreau might seem more acceptable to classroom philosophy if we decided that he provides an epistemology, one we could herald, with only a touch of tongue-in-cheek, as a "Theory of Exuberant Knowledge." Let's say we take the idea of sympathy with intelligence—being intelligently sympathetic and sympathetic to the intelligence or translucent intelligibility of things—as a warning against dry learning or mere argument or haughty neglect of the particular. As Thoreau puts it in his *Journals*, knowledge must be visceral, facts that are "warm, moist, incarnated." "A man has not seen a thing who has not felt it."[7]

For good or ill, Thoreau's scale of attention is geared to the particular, and he had little hope for abstract argument (whose meager glory redounds to the ingenuity and brilliance of the debater, rather than to the radiance of the matter at hand). I'm sure he'd think that hours watching seals at play would do more for appreciation of all life than arguments about why all life—not just human life—is precious, and that hours watching the sand cut thaw would do more for appreciation of all life than arguments about why the earth is alive. Here is how exuberant knowledge might work:

> As knower, you gain from the creature known a property you could not have had without the known's being what it is which is not merely your delightedly reflexive intimacy with it, but also your awareness of yourself as delightedly intimate in just that way. Your knowledge of any creature is therefore a kind of participation in it, as its being known by you a participation of it in you.[8]

This is knowledge attendant on vibrant participation in creation among fellow creatures in the modes of intimacy, mutuality, and delight.

Mourning

We're not the rational animal, the language-using animal, or the political animal (or not only these), but the wounded, grieving animal—mourning inevitable loss. For Thoreau, mourning is perpetual, interminable—not a fix, or a path to closure. Neither is it laced with dejection. Thoreau refuses the high drama of melancholy and lets the crack of dawn carry over and through his sighs of lamentation.

There's nothing dejected in Thoreau raising a cairn to Margaret Fuller by the surf off Fire Island where she drowned and was ravaged by sharks. He's on the trail of lost bones of his friend. Here is the passage from *Cape Cod*:

> I expected that I should have to look very narrowly at the sand to find so small an object, but so completely smooth and bare was the beach . . . that when I was half a mile distant the insignificant stick or sliver which marked the spot looked like a broken spar in the sand. There lay the relics in a certain state, rendered perfectly inoffensive to both bodily and spiritual eye by the surrounding scenery, a slight inequality in the sweep of the shore. . . . It was as conspicuous on that sandy plain as if a generation had labored to pile up a cairn. . . . It reigned over the shore. That dead body possessed the shore as no living one could.[9]

At first he sees only a sliver, a splinter, which he sings into a spar, and then a stone cairn—bones become a saint's relics, and the bouldered tomb communes with the wash of the sea, which "rolls on as it rolled five thousand years ago."[10] Indeed, it's an interminable longing and lamentation, delivered not in tears but in dappled majesty, Kaddish prayed to the magnification of creation, not in funereal wails, but in sighs that are breaths of delight.

Yet this is not solitary meditation. Thoreau's writing is not about *him*, but community, poetry in action: he celebrates Margaret Fuller and John Brown for their good—for *all*. He takes graveyard notes to remind a community that there are *no* forgettable persons—*no one* is so insignificant as to be forgettable. From the darks of the pond, he drags up a giant snapper by the tail, to be examined and praised in full light. He loved the telling detail that would make of the dead more than a scatter of bones. As we listen and join him, we too are brought to life—loss is *part* of the mix, but not *all*.

Thoreau can mourn the great and small, the anonymous Irish dead at an immigrant shipwreck and the world-historical John Brown.[11] Thoreau can even roughhouse with his own death and burial. He climbs Mt. Greylock to the ramshackle Williams College observatory to take in the expanse of three states below (Massachusetts, Vermont, New York). At night he configures loose boards abandoned at the foot of the observatory into a makeshift coffin and crawls in, securing the topmost plank with a rock, entombed until morning.

Thoreau wants to explore the transitions from death to rebirth to death again—in every available medium. He pursues cycles of grieving and celebration, mourning and renewal, never succumbing to depression, defiance, or despair. He invites me, and the village, to a community reel that includes turtles, foxes, trees, and mists on the rivers.

Community

Solidarity, community, and the ceremonial underlie Thoreau's practice of mourning. It is ceremony as he chases a fox, raises a sacred cairn, entombs himself on Greylock for rebirth at first light. His writing-as-mourning is not geared to repairing private loss, but is offered in affirmation and acknowledgment of—and partial participation in—*universal* loss and repair.

Through writing, Thoreau participates in all-pervasive cycles of loss and regeneration and joins in the kind of ground-bass steady mourning we find throughout the *Iliad*, say, where all men rage and weep. The corollary is that grief and celebration are not *inner* psychological states—or not only inner. They are part of a psychology that is fully *embodied* and that is *socially embedded*. We find three-way dances of energy—psyche, soma, others. These multiple couplings and inter-animations are strange only to those tyrannized by a lingering Cartesianism that partitions mind from body and both sociality. In Thoreau, as I see it, we have threefold couplings of body-spirit-other, other-spirit-body, all further animated by powers of place.

Thoreau's vigilant attention to loss and mortality and the allure of place calls for appreciation and praise, as does his raveling of loss into the fabric of prose. He weaves with such melodious ease that we forget we are mourning, hurting . . . or at last, *not* hurting.

At Last

My thought as I reread Thoreau is that much art and writing (religious and otherwise) is a rescue of things from decay and dissolution and is, if not a theodicy, then a rescue of the world from abandonment, which rescue slightly meliorates the inevitable wounds and trauma we endure.[12] Good Friday gives us memories of death and resurrection. Passover lets us remember captivity and escape. We are not—as I think of it now—*simply* rational, political, passionate, speaking, or ethical animals. We are that, but we are more. Plain and simple we are *wounded* animals, stammering ones.[13] We know (as Coetzee's Helen Costello can't help but testify from a wounded heart), that we are creatures who know and can't forget pain. And if we are Thoreau, we also know that despite the wounds, we can't stop singing the life that persists, sometimes a life fully present, sometimes one alluringly around the bend—the horse and the hound, the dove bittersweet.

The idea of interminable mourning is powerful—especially if you notice that any given site of mourning spills out over previous and subsequent sites. Imagination can start with a quite specific painful event, but it hardly ever stays put. What's "interminable" is not a *moment* of trauma or pain (unnaturally extended to a point of chilling, pathological fixation). Mourning and regeneration are "interminable" because the initiating loss-and-recoil impinges in ever new and widening (then narrowing) arcs on subsequent and prior disruptions and loss. Such reverberating loss and recovery fill memory and imagination. It just happens to make up the fabric of *any life we can know . . . or live.*

Something poetic or musical stays with us that links the parts of his vision (shall we call it) together. Song—the pitch of the voice, the rhythm of walking—does the weaving. He does not have an all-time frozen *position* on this or that for us to surround and attack or to pocket as a simple creed. He has the poet's prerogative of switching momentum or register or mood as words unfold. This prerogative frees him from being vulnerable to the ill-considered charge of inconsistency, or lack of system—as if these were matters to lament or object to, regardless of circumstance or intent. At times a free-floating, almost whimsical remark has inestimable value in freeing intelligibility for another—apart from its systematic development. So the loose assemblage of "takes" on life and death, on mourning and celebration, on creatures and creation, on nature and civil life, have a recognizable drift and Thoreauvian signature, even as they resist systematic exposition of the sort we'd expect from accounts from Kant or Spinoza.

Reading and recounting the drift of Thoreau is more like reading and recounting the drift of Montaigne, Kierkegaard, or Nietzsche. The *aperçu* of the moment can carry great weight, even as contrary ones, *aperçu tensed against* these, all emerge in their time. Still more, they emerge even as other collateral observations and exhortations also emerge in their time—and so often hang beautifully together in the process. They hang together as the mix of a difficult, flexible, and mobile mind in a reality that is also flexing and mobile and full of contradictory impulses and movements. To my ear, this means that Thoreau gives a better, more truthful account of the difficult realities creatures face, and face creatively, than a more streamlined and systematic account of the sort that appeals to philosophers of explicit order—and cool objectivity. Difficult times do not exclude joy, exuberance, intelligence, or sympathy.

Creation and the Glory of Creatures

Janet Martin Soskice

Near the Colosseum in Rome stands the Basilica of San Clemente, a medieval church built upon the ruins of a fourth-century basilica. It is the apse of the medieval church that concerns us here, for it contains a Byzantine mosaic, dating from 1110 to 1130 C.E., depicting the Tree of Life or (it is the same thing to the mosaic artist) the cross of Christ. The crucified Christ rests peacefully on a dark cross, eyes shut, flanked by Mary the Mother of God and St. John. Standard so far; but less so are the twelve doves that adorn the cross: these are the apostles, also represented as rams flanking the Lamb of God in the border below. But the real energy of the image of the cross springs, literally and figuratively, from the side of Christ. A stream of blood falls down his pierced side to fill a fountain that overflows into four streams that are the rivers of paradise. From these deer are drinking ("As a deer longs for flowing streams so my soul longs for you, O God"; Psalm 42:1). The cross seems to stand planted in the fountain, but, while it is a cross, it is also a great tree. From the living water at its base grows a vine, an acanthus, which fills the golden space of the apse. The tendrils of the vine unfold in voluptuous array, each coiling around a mosaic vignette—in one a man tends cattle, in another a woman feeds chickens. The vines embrace wild nesting birds, many of whom feed their young, and a cleric (supposedly St. Augustine), who writes in a book.

It is an image of the new creation, not of heaven so much as this world made harmonious and whole. The Byzantine artist has provided a vision of salvation construed, as etymologically it should be, as *salus*: health, flourishing, and well-being. The space of the apse is finite, but these tendrils could extend without limit, making nests for all of life each in its particularity. I would like to keep this image in mind as I turn to the topic of *creatio ex nihilo*.

In an essay on creation in the Psalms, James Mays writes, somewhat acerbically, that "creation" in the sense as commonly used today is not a notion that appears in the Psalms:

> the term (creation) has come to mean no more than the natural world in the vocabulary of New Age religion, nature romanticism, environmental enthusiasm, artistic aestheticism, and even in unselfconscious traditional religion and liberal piety. There is no term or text in the Psalms, or indeed in the entire Bible, for creation in this sense.[1]

If slightly overdrawn, his point is well taken—creation in the Bible is about an activity and the outcome of the activity, and the "actor who creates is always Yahweh."[2] The doctrine of creation as found in the writings of the early Christians is also not about animals and birds. It is not about nature, at least not primarily, but about God. As in the Psalms, creation is both the activity of God and the outcome of God's activity. We need as modern readers to avoid an easy equivalence of nature and creation, not least because modern secular writers readily speak of creation when they have no intent of implying the existence of a Creator. Conceptually, creation implies a Creator, whereas nature does not. Indeed, some early modern writers preferred to use the term *nature* (suggesting a freestanding, autonomous whole) to get away from the idea of a Creator. By contrast and in reaction, the seventeenth-century rabbi David Nieto, leader of London's Sephardic community, insisted that Jews believed in divine creation and not in a universal nature.[3]

It should be added that the developed theology of creation does not come from the Book of Genesis, or not from Genesis alone. De-

spite their reference for the biblical texts, the rabbis and the theologians of the early church seemed uninterested about any story of
cosmic origins involving a literal six days (most in fact rejected this,
since the sun and the moon were only created on the fourth day, so,
they reckoned, the "days" could not be days in our sense). The guid-
ing citations are more frequently from the Psalms and the Book of
Isaiah, writings that repeatedly link God's power to create with his
power to save. Here are some examples:

> Our help is in the name of the LORD, who made heaven and
> earth. (Ps. 124:8)

> By the word of the LORD the heavens were made.
> And all their host by the breath of his mouth. . . .
> Let all the earth fear the Lord;
>   let all the inhabitants of the world stand in awe of him.
> For he spoke, and it come to be;
>   he commanded, and it stood firm. (Ps. 33:6–9)

> Praise him, sun and moon;
>   Praise him, all you shining stars!
> Praise him, you highest heavens and you waters above the
> heavens!
> Let them praise the name of the LORD,
>   for he commanded and there were created. (Ps. 148:3–5)

> Shower, O heavens, from above,
>   and let the skies rain down righteousness;
> Let the earth open, that salvation may spring up,
> and let it cause righteousness to sprout up also:
> I the LORD have created it. (Isa. 45:8)

It is worth noting that the Psalms and Isaiah are the biblical texts
most frequently cited by both Qumran and by the New Testament
writings.

We know that the period of Second Temple Judaism was a time of upheaval and distress for the Jews. Foreign domination coincided with a foment of Messianic expectation. There were debates about the coming of the Messiah and the faithfulness of God to creation, and here the Psalms were perceived to be to the point. It is not surprising that texts from intertestamental literature contain what seems to be the first clear expression of the doctrine of *creatio ex nihilo*, which I will define, for present purposes, as the teaching that God has created all that is, including matter, space, and time, and does so freely out of love.

In 2 Maccabbees (a text hostile to Hellenism but written in excellent Greek) the mother of the seven martyred sons encourages them to lay down their lives, saying,

> I do not know how you came into being in my womb. It was not I who gave you life and breath, nor I who set in order the elements within each of you. Therefore the Creator of the world, who shaped the beginning of humankind and devised the origin of all things, will in his mercy give life and breath back to you again, since you now forget yourselves for the sake of his laws. (2 Macc. 7:22–23)

From a people in tribulation comes texts that speak of deliverance, of a God who created and who can recreate, a God who gave life and who can restore life. These are also the first of the Jewish writings where we see consideration of the resurrection of the dead, a question still debated by the Sadducees and the Pharisees in the time of Jesus. In the same story from 2 Maccabees the mother of the youngest brother, last to go to his death, leans toward him and says in their native language (presumably Aramaic),

> I beg you, my child, to look at the heaven and the earth, consider and see everything that is in them, and recognize that God did not make them out of things that existed. And in the same way the human race came into being. Do not fear this

butcher, but prove worthy of your brothers. Accept death, so that in God's mercy I may get you back again along with your brothers. (2 Macc.7:28–29)

The implication is that the God who made "all that is" can surely raise the dead, an argument found later in Islam. Resurrection involves the power to give life, to restore. In the books of the New Testament resurrection is linked to the promise of a new creation that, like the first, is made by God's Word.

We have, then, in 2 Maccabees, and in the Psalms and Isaiah, powerful reflections on creation. Whether, however, this is *creatio ex nihilo* will depend on how that teaching is defined. A good deal of the debate on "how early?" and "how Jewish?" is that *creatio ex nihilo* has revolved around the question of matter—did God create it? When was it insisted that God created matter? And how much did the ancient sages really worry about the question?[4] The first chapters of Genesis could be read, and were read, by some Jewish and Christian writers in antiquity as suggesting God created out of "something" that already existed—some primordial element or out of matter (*ex hyles*). Interpretation of the Hebrew *tohu wa-bohu* of Genesis 1–2 was, and remains, contested.[5]

Another intertestamental text, the Wisdom of Solomon, speaks of God's "all powerful hand, which created the world out of formless matter" (Wisd. of Sol. 11:17). Does this suggest *creation ex hyles*, or, alternatively, since it was written in Greek, might it be a gloss on the Septuagint's rendering of *tohu wa bohu* with two Greek alpha-privatives, "And the earth was invisible and unstructured"? It certainly seemed open to Jews reading Hebrew scriptures, and to Jews and Christians reading the Greek of the Septuagint, to see Genesis describing God as working creating from "something," a view especially congenial for readers versed in Plato's *Timaeus*, where the demiurge molds a preexisting matter. Enough early Christian thinkers may have been drawn to the Platonic notion that God molds creation out of preexisting elements for Origen (c. 185–284) to have expressed himself with exasperation against this view in his *First Principles*.[6] However, the most reasonable conclusion to draw about the ambiguity of *from nothing / from something* in Hellenistic Jewish texts of this period is that this was just not something that had begun to trouble them.[7] It will become a point of pain for Christians in the second century when countering Gnostic suggestions that matter was evil, recalcitrant, or outside the purposes of God.

What we can say with certainty is that from early on the Jews believed in a God who was a free and loving Creator and that this distinguishes their accounts of origin from those of Plato and Aristotle.[8] The earliest theology of creation seems to be, like the language of the Psalms, confessional and doxological, underscoring the freedom, power, and goodness of God. Such is still the tone of the New Testament and of the first-century Christian fathers, for whom belief in God, Creator of heaven and earth, is a *central confession*, as it would be in the Creeds.[9]

The striking and distinctive teaching we call *creatio ex nihilo* has many strands and was subject to many developments, from the intertestamental period through to the Middle Ages and the decisive writings of Moses Maimonides and Thomas Aquinas. It is not a teaching of Greek philosophy. Aristotle thought it absurd: from nothing, nothing can come. In Aristotle's view God and the universe were both everlasting and corollaries of each other. There could not be the universe without God, nor could there be God without the universe. In Jewish and Christian teaching, by contrast, God need not have created at all. Aristotle's God is the cause of motion, but not the cause of beings. Plato's demiurge molds a preexisting matter. That God made "all that is" and did so from no compulsion but freely out of love seems to emerge distinctly from Judaism and early Christianity in defense of a conception of God who has made and cares for all things. It is a rupture in ancient cosmology. The Creator, moreover, is called upon as Savior in time of distress.

> Happy are those whose help is in the God of Jacob,
> whose hope is in the LORD their God,

who made heaven and earth,
the sea, and all that is in them;
who keeps faith forever;
who executes justice for the oppressed;
who gives food to the hungry. (Ps. 146:5–7)

One who creates can redeem. The Psalms repeatedly link God's creative and saving power to the story of the Exodus and the crossing of the Red Sea.

Gerhard May, the doyenne of students of *creatio ex nihilo*, insists that this teaching is really only fully established in Christian writings of the second century, driven by controversies with Gnosticism. May's analysis, however, revolves around the question of the creation of matter, only one of the aspects of *creatio ex nihilo*.[10] If, as it seems on May's own analysis, the question of where matter came from didn't much concern earlier thinkers, it might be better to look to alternatives. If we make our focus the sovereignty and freedom of God in creating and the idea that all creatures have their being by God, we might find clues to another prehistory for the teaching. And here Philo makes an interesting study.

Philo is a Greek-speaking Jew of first-century Alexandria, a contemporary of Jesus and Paul but with no apparent knowledge of them or the infant Christian movement. There had been significant Jewish populations in Alexandria, probably since its foundation as a "new town" by Alexander the Great in 331 B.C., and they were (like all the city's inhabitants) Greek-speaking. It is thought that many of the books of the Hebrew scriptures were translated into Greek (becoming what we call the Septuagint) for the needs of this Alexandrian Jewish population. Philo is wealthy and well-educated in Greek philosophy and, like some of the earliest Christian writers, feels little tension between his Jewish faith and the Greek philosophical writings he has been brought up in. In fact, he thinks Plato is correct on many points, not least because Plato has, through some obscure means, learned from Moses.[11] He knows Plato's *Timaeus* and invokes it in his own commentary on the Book of Genesis.

However, Philo is not in his own terms a philosopher. He writes as an exegete of the Jewish scriptures, which he, like St. Paul, reads in Greek. His writings are all extended commentaries on scripture, heavily informed by Hellenistic thought.[12] Philo is interesting to us because we see in play in his writings all the principles central to the doctrine of divine transcendence that is enshrined by *creatio ex nihilo*.

Philo is widely held to either not hold to *creatio ex nihilo* or to be inconsistent on it, especially when it comes to the status of matter. Gerhard May thinks Philo "postulates a pre-existent matter alongside God," molding a matter that verges on the malign.[13] On May's reading, Philo, though not a dualist who posits matter as "an ontologically equal principle alongside God," nonetheless "took over the Greek teaching about pre-existent matter, without thinking it through independently."[14] The picture is of someone who has not yet absorbed the implications of a particular received (Platonic) belief for the teaching about the sovereignty of God to which he is committed.

But if the doctrine of creation was in flux in this time, then so too was Greek philosophy. As already mentioned, neither Plato nor Aristotle conceived of a Creator God. For instance, in Aristotle's view there could be no world without God, but equally no God without the world. This changed in the period we know as "middle" or "eclectic" Platonism, the period during which Philo and the New Testament authors wrote. A varied mixture of Stoic, Aristotelian, and neo-Pythagorean ideas led Platonists of this period to posit a belief in a One who was the source of the many, a position much more amenable to Jews and Christians, and most fully worked out by the third-century neo-Platonist Plotinus.

Philo can be called a middle Platonist, for such is his philosophical formation, but only at the risk of thinking of him principally as a philosopher. In his own terms Philo was an exegete of scriptures. It was always scripture that was his primary authority, and he was jealous for what he believed scriptures told him of God. So when John Dillon writes that all Philo brings to the Middle Platonic mix "is a distinctive streak of Jewish piety, a greater personal reverence for God than one would expect to find in a Greek philosopher" . . . (which) "also leads on occasion to a downgrading of the ability of the human intellect (unaided by God's grace) to comprehend truth," our ears should prick up, for, as I shall argue, it is this "greater reverence for God," which we might understand as a greatly enhanced sense of divine transcendence, that will prove so important.[15]

Philo wrote a commentary, known to us as *De Opficio Mundi* (*On the Creation of the Cosmos*), on the first books of Genesis. In these he posits, as do other middle Platonists, a double creation in which God creates an "intelligible" cosmos that is the model for the "sense-perceptible" cosmos. Things are further complicated by his insistence that the "intelligible" cosmos is not chronologically prior to the "sense-perceptible" one, since God does not create in time. Philo believes creation is instantaneous and the "days" of creation are a figurative way in which Moses, whom he takes to be the author of Genesis, speaks.[16] Even in this work Philo seems not so much interested in physical cosmogony (far less in anything approaching modern cosmology) than in laying down a picture of God as Creator or the world and its order. The ordering of the "days" of Genesis he understands as a figurative means (since creation is "simultaneous") by which Moses anticipates the "ordering" given by the Law at Sinai. In this work Philo persistently stresses that God as Creator is one, and that God alone creates.[17]

The expressions that border on *creatio ex nihilo* are voiced in Philo, more often than not, when his topic is the sovereignty of God and thus often in contrast to the philosophers: "God, being One, is alone and unique, and like God there is nothing" (*Leg.*, II.1). Here are just a few more of Philo's views: God has created the world out of nonexistence. All things are dependent on God, but God is sufficient to himself and was so before the creation of the world.[18] He creates from his goodness and governs what he creates. He does not change or alter. God, says Philo, "created space and place coincidentally with the material world" (*Conf.*, 136). God created time itself, "For there was no time before the cosmos, but rather it either came into existence together with the cosmos or after it (*De Opficio Mundi*, §26).[19] In concluding *On the Creation of the Cosmos according to Moses*, he attacks Aristotle, saying that one of the beautiful lessons taught us by Moses is that "the cosmos has come into existence," a teaching especially important "on account of those who think it is ungenerated and eternal, attributing no superiority to God" (*De Opficio Mundi*, §71; see also §7).

My own interest in Philo came first not from what he had to say about God as Creator but what he had to say about naming God. This turned out to revolve entirely around his elevated ideas of divine transcendence. For Philo, God is, strictly speaking, "unnameable," and indeed Philo is our earliest source for a number of divine epithets that relate to this. John Dillon says, "At *Somn.*, 1:67, for example, God is described as 'unnameable' (*akatanomastos*) and 'unutterable' (*arrhetos*) and incomprehensible under any form (*kata pasas ideas akataleptos*), none of which terms are applied to God before his time by any surviving source."[20]

Since God is Creator and not a creature, God cannot strictly *be like* anything created as a being. It follows that we cannot *class* God or insert God into any category appropriate to our created kind. On the other hand, and equally from his Jewish faith, Philo knows that God must not be left in the realms of abstraction. The God of the Pentateuch is personal and providential. In attempting to bridge this gap between the God who is totally *unlike* all created, finite beings and yet who is nonetheless present to us in the intimacy of disclosure, Philo anticipates a problem central to what is sometimes called mystical or negative theology but that I prefer to call the theology of the divine names. For if we say that God is "wholly Other," we must explain what justifies our language of praise and adoration. Jews (and Christians) do not characteristically want to say nothing at all about God. The God of the Pentateuch is personal and providential. How can such a way of speaking be justified?

Whenever, in his commentaries, questions about knowing and naming God arise, Philo turns to the example of Moses. He returns again and again to Exodus 3, the burning bush, to Exodus 20, where

Moses ascends the mountain of Sinai and "approaches the dark cloud where God was," and to Exodus 33, where Moses asks to see God's glory, all texts of epistemological salience for Philo. Indeed Exodus 3 and Exodus 33 receive prominence not just in Philo but among the rabbis and early Christian theologians as instances of Divine self-naming.

Philo, writing in Greek, routinely understands the "I Am Who I Am" of Exodus 3 metaphysically—God, for Philo, is the Existent (to on). It is not, says Philo in an important expansion on the text, "the nature of Him that IS to be spoken of, but simply to be." We need to remember that Philo's exegesis, even when, as here, it sounds most metaphysical, is driven by his understanding of who God is in his scriptures, the Creator of all that is. Here is another of his comments on Exodus 3:

> "I am He that IS" (Exod.3:14), which is equivalent to "My nature is to be, not to be spoken." Yet that the human race should not totally lack a title to give to the supreme goodness He allows them to use by licence of language, as though it were His proper name, the title of Lord God of the three natural orders, teaching, perfection, practice, which are symbolised in the records as Abraham, Isaac and Jacob. (De. Mut., II.12–13)

A God who creates heaven and earth cannot properly be named in earthly words; however, "by licence of language" that same God has given names us by which God may be named. The name Philo believes is given to Moses in Exodus 3 is, curiously, "the Lord, the God of Abraham, and Isaac and Jacob." This, Philo believes, is the name that is the name for all generations, which indeed incorporates the "generations" of the people of Israel.[21]

To recap, the doctrine of creation *ex nihilo* is a biblically inspired piece of metaphysics—not a teaching of Hellenistic philosophy pure and simple, but something that arises from what Greek-speaking Jews found in their scriptures. Because God is the "creator of all that is" and not a creature, the divine essence is, strictly speaking, unknowable to us. It follows that God is unnameable in any way that might suggest a delineation of the divine essence. This philosophical and rational constraint paradoxically means that primacy in matters of naming God must be given to revelation and to scripture. Indeed apparent philosophical inconsistency on the question of the preexistence of matter in Philo and the early fathers may arise because they are driven not by philosophical premises and concern for cosmological consistency but by the scriptures that they are exegeting. So, for instance, Clement of Alexandria (late second century) will mix Hosea with Isaiah, Psalms, and Jeremiah to extol the God who is not far off and who fills heaven and earth (Jer.23:23). How this "filling" might be parsed philosophically is not his primary concern. But it becomes a concern for Christian writers who must explain how their God is not an adjunct of the natural world, but the one from whom all things have their being and on whom every thing in its creaturely reality entirely depends.

I have avoided thus far such terms as "negative theology" and "cataphatic" and "apophatic" as anachronistic to this early stage and because I am puzzled over who is and who is not considered to be a practitioner of "negative theology"—usually not Augustine, for instance. But here is Augustine in the *De Trinitate*:

> we should understand God, if we can and as far as we can, to be good without quality, great without quantity, creative without need or necessity, presiding without position, holding all things together without possession, wholly everywhere without place, everlasting without time, without any change in himself making changeable things, and undergoing nothing. Whoever thinks of God like that may not yet be able to discover altogether what he is, but is at least piously on his guard against thinking about him anything that he is not.[22]

Augustine never tires of saying that what God "is" we shall never know, at least not in this life. These negations are not the collapse of reason in despair but rather its proper deployment in the face of the

One who is the source and cause of all that is. By Augustine's time this is firmly linked in Christian thought to *creation ex nihilo*.

As with Philo, it is the doctrine of *creatio ex nihilo* that lies behind this crisis in speech. Indeed, as Carol Harrison has persuasively argued, it is this teaching and not random sampling of Plotinus that delivered Augustine from the materialism that still held sway in his thought when he was a follower of the Manicheans.[23] It is this developed sense of divine transcendence that enables him to see that God is at once entirely beyond our understanding and yet most intimately present—closer to me than I am to myself. It is that which enables him to affirm that "all that is" is Good and that God can never be thought of as a spiritual overlord of a debased material realm. God, for Augustine, is Being Itself and the source of all being. In this teaching, "souls" and "angels" are just as much creatures as dung beetles and dinosaurs. This point is expressed in the Nicene Creed by confessing God to be "Maker of heaven and earth, and of all things visible and invisible."[24]

If Augustine's *Confessions* shows the same dynamic with regard to knowing and naming God as Philo, then one would expect soon to find Moses and the "I AM WHO I AM" of Exodus 3—and we do. Shortly after his conversion, and while with his friends at Cassiciacum, Augustine reads the Psalms. The final version of Psalm 4, he tells us, wrung a cry from the very depths of his heart. Here is his full comment in the translation of Maria Boulding:

> *In peace! Oh. In Being itself!* What did it say? *I will rest and fall asleep.* Yes, who shall make war against us when that promise of scripture is fulfilled, *Death is swallowed up in victory?* In truth you are Being itself, unchangeable, and in you is found the rest that is mindful no more of its labours. (*Conf.*, IX.11)[25]

For many modern theologians to answer the question "Who are you, God?" with "Being Itself" is worse than no answer at all—the Christian theologian trading divine insight for dubious Greek metaphysics. But this is to misunderstand Augustine, for, as Maria

Boulding has pointed out, *idipsum*, "the Selfsame," is linked by Augustine here and throughout his works with the mysterious name for God as given to Moses from the burning bush—the "I AM WHO I AM." Augustine may well have found the notion that God is "being itself" in Porphyry, the student of Plotinus, some of whose works he read in Milan, but he need not have done. The idea was freely available in the works of earlier, Christian middle Platonists, and embryonically, at least, in Philo, whose work was well known to Augustine's mentor St. Ambrose.[26]

As mentioned, we do not find a creator God in either Plato or Aristotle. We do find a One that is the source of the Many in Eudorus, an Alexandrian philosopher of the first century B.C.E. But why should we not think Eudorus already influenced by Hellenistic Judaism, given the large and literate Jewish population of his Alexandria? Certainly by the time we get to Porphyry and Plotinus in the late third century, we must take into account Numenius of Apamea, a second-century "middle" Platonist and acknowledged influence on Plotinus. Numenius has a Platonic view with a First God and a Demiurge—the First God he calls "Being Itself" (*ho on*). This goes far beyond Plato, who makes no such claim for God. What is remarkable is that Numenius was self-confessedly interested in and impressed by the Jewish writings, and it is these, and especially the Septuagint of Exodus 3:15, that seem the likely source of his name for the First God ("Being Itself"—*ho on*). Numenius credits Moses with receiving the "revelation" that the First God is "Being."[27] Even were we to leave Numenius out of the picture it is now widely accepted that Plotinus knew the works of Philo, who, as we have already seen, already in the first century, spoke of God as "the Existent" and did so on the basis of Exodus 3 and the request of Moses for a name.

Nor can this identification be simply attributed to Philo's Hellenism. The Palestinian Targums, Pseudo-Jonathan, written in Aramaic, glosses the "*ehyeh esher ehyeh*" (I Am Who I Am) of Exodus 3:15: "He who spoke and the world came into being, spoke, and everything came into being."[28] This linking of the "I AM WHO I AM" of Exodus to the Creator of the book of Genesis is the more interesting, since the Targumist is likely to have realized, as Philo and most of the Christian fathers did not, that the Hebrew "I AM WHO I AM" (*"ehyeh esher ehyeh"*) is a pun or play on words on the Tetragrammaton (YHWH), which gives the appearance of a form of the verb "to be."

I do not think we should complain that the wholesome streams of Judaism and Christianity are here contaminated by Hellenistic philosophy. This is a discredited exercise, and Robert Wilken is surely right in saying that we should speak now not about the Hellenization of Christianity but the Christianization of Hellenism[29]—and nowhere more so than in the distinctively Jewish and Christian teaching of *creatio ex nihilo* and the reflection on Being and participation attendant on it.

Judaism and Christianity precipitated a revolution in Western metaphysics. *Creatio ex nihilo* excludes both an Aristotelian cosmology, in which the world and God necessarily coexist and imply one another, and neo-Platonic emanationism, in which the world flows unconsciously and ineluctably from the Godhead like rays of light from the sun. Creation for Jews, Christians, and (when they appeared on the scene) for Muslims is intentional. God wills to create. It is a gift. Also ruled out is the notion of a "descending hierarchy of existence" in which the more "material" a being is, the further it is from God. The distinction, as David Burrell often reminds us, is between the Creator and everything else.

It may appear that these metaphysical reflections are remote from the array of living creatures that unfold in all their mosaic "thisness," their place and particularity, in the San Clemente apse, but I would suggest they are quite the contrary.

In the Septuagint version of the famous "fiery furnace" story of the Book of Daniel (purged from modern Protestant translations but still to be found in Catholic Bibles) the three servants of God, Shadrach, Meschach, and Abednego, thrown into the flames by King Nebuchadnezzar, begin to sing, glorifying and blessing God and inviting "all things the LORD has made" to bless the LORD: "Angels of the Lord! all bless the Lord," heavens, waters, sun and moon, showers and dews, fire and heat, night and day, frost and cold, lightning and clouds, mountains and hills, springs of water, seas and rivers, sea beasts, birds of heaven, animals wild and tame, sons of men, servants of the Lord, all are enjoined to bless the Lord, for "he has snatched us from the underworld, saved us from the hand of death . . . for his love is everlasting" (Dan. 3:51–90). This is a God of creation, salvation, and justice, and with the Christian conversion of the "Lord who saves" into Jesus Christ, it is the God of the new creation depicted in the San Clemente apse.

Creation is not something that happened to the universe long ago. It is not the distant accomplishment of a distant God. *Creatio ex nihilo* underscores the belief that God imparts the being of all created things, visible and invisible. The world is graced in its createdness, which is happening all the time. In Christian understanding it is the Word through whom all things are made who redeems and renews all things—hinds and stags drinking from the waters of life—the new creation that embraces in its tendrils the nesting birds, the sheep and the shepherd, the sober bishop at his writings.

Care of the Soil, Care of the Self: Creation and Creativity in the American Suburbs

T. Wilson Dickinson

> We need to be clear about how changing concrete and social conditions mitigate or promote our capacity for attention, care, and responsibility.
>
> —*Norman Wirzba*, The Paradise of God:
> Renewing Religion In an Ecological Age

> Profound, creative, grace-filled spiritualities produce grace-filled environments; banal, impoverished, alienated spiritualities produce alienating environments. If that is the case, and I believe it is undeniable, then theology is anything but incidental to the debate about the built environment which is such a vital dimension of the human future.
>
> —*T. J. Gorringe*, A Theology of the Built Environment:
> Justice, Empowerment, Redemption

The smell of freshly cut grass caught on the breeze carries, for me, largely positive associations. Indicative of summer—a time of relaxation and frequent celebration—the first hints of this aroma after a season of cold and solitude are typically welcome. As I have been among the numbers of the itinerant intelligentsia for quite a while, it has been some time since the activity of mowing has been my responsibility. When that changed this past summer, and I found myself pushing a strange, unpleasantly loud contraption around the yard, different associations were brought to mind. As the sweat from my brow reached my lips, like a bit of madeleine, I was taken back to different memories, to afternoons in my adolescence when mowing was one of the few activities that I could use to translate my labor into money.

These memories and alternative associations led me to question the purpose of the activity in which I was taking part. I am not sure if I share in the aesthetic values implicit in such an activity. The house next to our previous residence was vacant and in profound disrepair. It was owned by a bank, and the yard was left untended. The result was not lamentable; it was beautiful. The grass grew up to create an idyllic meadow, and little blue flowers were scattered across its length and breadth. My wife and I followed suit with the small patch of greenery behind our apartment. While this mostly just yielded a thicket of the largest dandelions I had ever seen, these weeds served as the perfect food and cover for a group of golden finches that would frequently visit.

My abiding question as regards this practice is not its purpose but its effect. I wonder what sort of relationship this monotonous gas-powered routine fosters between its practitioners and their habitats. What sort of community does this carefully maintained landscape create? Does the suspended state of plant life that surrounds those who live in the American suburbs bear witness to a collective anxiety about disorder and change? Does the monoculture of the expansive green grass betray a social condition that is uneasy with difference, disorder, and diversity? Might the existential conditions of the suburbs be characterized as both a cause of the looming ecological crisis and an obstacle to addressing it?

In what follows I hope to draw attention to the habitat that surrounds so many of us who live in North America, so that we can see it again. So that we can see the familiar places and habituated practices that are invisible insofar as they are familiar and that often serve to cover over and obscure the contours of the places in which we live, the larger ecological connections that our everyday activity has, and the humbling and dynamic relationship that exists between creativity and creation. Doing so will demand that theological inquiry attend to the significance of place and the built environment, while also recognizing the reciprocal relationship that these have with formative practices.

Caring for Philosophical Theology

This effort to reexamine the existing relationships between social, religious, and ecological orders finds a strong precedent in the Hebrew Prophets. As Ellen Davis has helped to highlight, the botanical and agricultural imagery that leaps from nearly every page of scripture need not be seen simply as metaphorical artifice, but as grounded in ancient agrarian outlooks that presuppose a deep connection between theological commitment and the caretaking of creation. The objections that the Prophets waged against the political and religious elite were concerned with the abuse and the lack of care for a *land*, for an entire habitat, and not just the mismanagement of a *nation*.[1] On Amos's account the rulers of Israel have not simply exploited the rural poor, but have disrupted their relationship with the soil and the covenant with God.[2] As Amos suggests, there is a connection between the arrogance, selfishness, and opulence of the elite and the devouring locusts that emerge from the king's freshly mown grass (Amos 6:4–6; 7:1).

By drawing this creative and imaginative connection, Amos appeals to the heart (*lēb*) of his readers, which Davis tells us "in biblical physiology [is] the organ of perception and response."[3] Amos seeks to challenge and reeducate their habits of thinking and seeing so as to transform their everyday relationships with each other, creation, and God. In this way he creatively challenges a hubristic and habituated vision of human creativity in order to transform his hearers' relationship with creation.

The imagination that is being appealed to in this prophetic moment is not simply something that is fantastical. These imagistic connections are significant because they are embedded in what can

be called a social imaginary. Charles Taylor defines the social imaginary as "the ways people imagine their social existence" through the
interconnected web of ideal images, stories, expectations for normal
behavior, and common practices.[4] Reciprocal relationships are built
up between exemplary images, conceptual formulations, and repeated practices. The engagement with the imaginary does not simply take place in the mind or in the formulation of the text, but is
also embodied in everyday activity. My concern with the quotidian
practice of mowing the lawn, then, is not simply with the doctrine
or idea that it ultimately expresses about creation; I am interested in
a more holistic account of the way of life and forms of relationship
that it cultivates or undermines. This concern is doctrinal, imaginative, and practical all at once. Treating religious activities and places
in merely symbolic terms can often serve to obscure the degree to
which they—along with ideas—are formative of capacities, communities, relationships, and habitats.

To address, and perhaps even transform, these forms or ways of
life, I turn to a theological approach stretching back to early Christian philosophers that combined prophetic vision with the Hellenic
practices of the care of the self. In the broadest terms, the care of
the self was, as Michel Foucault explains, a "way of considering
things, of behaving in the world, undertaking actions, and having
relations with people." It could also be characterized as a "form of
attention, of looking." More specifically, this care was often carried
out through techniques or practices—like meditation, examining
one's conscience, or considering the name, components, and functions of things.[5] The care of the self, then, was a way to engage with
or transform the practices, images, and expectations to which one is
habituated through the social imaginary. These spiritual exercises,
as Pierre Hadot called them, served to shape, purify, and transform
the self.

While often focused on gathering or transcending the self, these
exercises of care can also be given a more relational valence. Jean-
Louis Chrétien characterizes prayer in this way, as it is always addressed to another. This is not the act of an autonomous self or of

an atomized individual, but one that is only opened in relationship.[6] Speaking more formally, Martin Heidegger characterizes care in similarly relational terms. He does not speak of a self or a subject, but a "being-there," which is always already "being-with" a world of others and concernful things.[7] Michael Fishbane adds that caring for the self in a mindful, rather than mindless, way shows "our care for the world; for this kind of philosophical attitude opens the world to reflective wonder, and such thoughtful wondering promotes an ongoing sense of duty and responsibility."[8]

Interestingly, when Heidegger is introducing these connections he speaks of the setting with which I began, a patch of land. He notes that we are not simply surrounded by objects, or even tools, but by a world that emerges out of the care of others. He writes, "The field, for example, along which we walk 'outside' shows itself as belonging to such and such a person who keeps it in good order."[9] While Heidegger simply treats the field as an example, Norman Wirzba proposes that the care of the soil has much deeper roots in the ways that we live and relate—that is, in our culture. Though the word *culture* came to refer to something that was quite apart from nature around the sixteenth century, the middle English usage of the word referred to a "cultivated piece of land, a field upon which people worked to secure their livelihood."[10] Culture understood in this manner is not a human refinement outside of or beyond the natural order; it is about the way of life that emerges from a close relationship with the contours of the place in which one lives.

If we follow Wirzba's further suggestion that we hear the faint echo of the Latin *cultus* in our term *culture*—evoking practices of cultivation and care—then perhaps the activities of philosophical theology will not seem so distant from everyday life. Debates in philosophical theology might not simply be concerned with conceptual formulations, but might seek to address the practices that shape our lives and our capacity to attend to the relationships that constitute our world. This would mean that the exercises of care that frame the pursuit of wisdom might not be so far from the activities through which one cares for the soil. Therefore, to consider the character of creation, it might be necessary to examine the practices that are formative of character and vice versa.

The Pedagogy of the Lawn

Far from making its inhabitants attentive to or knowledgeable about their surroundings, the practices of suburban life often seem to cultivate ignorance, disregard, and even contempt for the particularities and limitations of a place. This is reflected not simply in the strategic leveling of mountains for highways, houses, and coal, but also in the daily tactical acts of commuting, shopping at supermarkets, and mowing. The suburban American lawn follows a similar design as it demands a homogenous vegetative space, regardless of context, and constant maintenance from its inhabitants. Green lush lawns do not simply sprout up, but are maintained by the uniform and thoughtless application of fertilizers, pesticides, herbicides, machinery, and irrigation systems.[11] As a result, this sprawling suburban carpet does not simply destroy the preexisting ecosystems it serves to replace, it is also unsustainable—as its chemical inputs pollute the very water sources that its irrigation depletes.

Deeply connected to this environmental damage is the manner in which the lawn can shape suburbanites' expectations of and outlook on creation as a domain that can simply be bent to our will and desires. It is a space that confirms the consumer's assumption that, as Wirzba warns, "the earth, combined with our own technological ingenuity, will always be able to provide for these wants."[12] Erasing these limits and severing the obvious ties of dependence, these practices might serve to promulgate fantasies of autonomy.

The problem of the cultivation of the lawn and a culture of hubris cannot be reduced to a fault of individual choice or vice. To diagnose the dilemma in this way would be to grant the arrogant vision of human agency and ecological receptivity that it ostensibly promotes. Whereas the lawn and consumer culture might trade in myths of autonomy and images of unlimited human creativity, the lawn is formed by subjects who are subject to a very rigid social order and a demanding, if bizarre, manifestation of creation. The sins of this degradation of soil and spirit are social. The lawn, after all, is not typically a space of great individual expression, but one of conformity and social pressure. Paul Robbins, citing a 2001 survey, notes that suburban homeowners who use chemicals on their lawn are *more likely* to both be motivated by a sense of social "well-being" and to be aware that such activities are harmful to local water sources.[13] In this scenario the public goods of property values and social cohesion outweigh the costs of ecological destruction, not to mention the demands of money and labor.[14]

These demands for conformity betray the ruse of autonomy that is offered by consumerism, in which the supermarket provides an infinite variety of foods from which to choose, each of which is derived from the same twisted and unhealthy corn products. The scene of the supermarket, like that playing out on the lawn, is not simply the performance of the individual exercising his or her creative might on creation. The person who mows his lawn is not simply enacting or choosing a privately conceived wish. He is responding to complex communal pressures (some of which are a matter of economic coercion in the form of advertisements or political demands in the form of legal restrictions) and the biological needs of their turf grass habitat. The American lawn is not the eradication of organic life, but cultivates relationships with a particular form of plant life and a specific kind of habitat—one in which both creation and human creativity seem to be stunted in the midst of broken networks and relationships.

The lawn, then, is not simply the expression of individual will or even a symbol of an underlying cultural idea, but a space of reciprocal formation and transformation. The lawn and the practices of cultivation that accompany it are laden with implicit values and ideas, while they also serve to form tangible and embodied networks and relationships. It might be helpful, then, to think of the lawn as a school—steeped in inheritance, kept alive by practices of formation, and oriented by visions of goodness and hope. This brings out the active and communal aspect of the teaching or *doctrina* of creation.

The pedagogy of mowing the lawn resembles Paulo Freire's critique of the educational model of lecturing and testing that trains students to passively "receive, memorize, and repeat" information. In this environment students learn about their world in terms that are static and established.[15] On the one hand this gives them the impression that creaturely knowledge exerts a power that seems to exceed all limits, but on the other it places them in an entirely passive relationship with that world. Rather than coming to know the vulnerable and particular manner in which vegetation and knowledge grows, the repeated and limited activity of maintaining the same patterns and structures serves to limit both one's vision of creation and one's creative capacity.[16]

The lawn is not educational simply for those who maintain it, but perhaps also exerts a formative influence upon the external and internal understanding of those for whom it serves as a habitat. Accepting only one size, seeking uniformity of color, and caught in the same stage of life, the lawn seems to reflect a homogenous cultural architecture that is unable to care for or attend to the particularity of spaces, the differences of others, or the limitations of our capacities. As Wirzba notes, the uniformity of the suburbs also makes those within it blind to the needs that exist outside its borders, as the "world of manicured yards conceals the blights of poverty, land degradation, and economic injustice."[17]

Might the lawn, then, be both the training ground and the icon of a social order that is unable to deal with any significant difference or dependence that does not conform to its order? What is often most objectionable in the suburbs, or rather the discontent that is most easily expressed and heard, is that which concerns disorder. The disorders of unkempt lawns, criminal behavior, or demographic anomalies that might bring property values down are the easily articulated subjects of concern. Such fears are consistent with Catherine Keller's characterization of the colonial episteme in which "evil is always disorder rather than unjust order."[18] The repetitive and compulsive activity of mowing could be understood not simply as the expression of a certain kind of order, but as the manifestation of anxiety about disorder or chaos. As Keller later observes, "Fear transmutes into phobia when it obsessively repeats itself, coding its dread and loathing in a symbolism that may in fact make it more difficult to face real threats."[19] In the suburbs, the obsessive and repeated movements of the lawnmower produce a symbol and a school that does just this.

The Garden School

In contrast to spiritual exercises or forms of cultivation that lay a homogenous and overly controlled order on simultaneously ecological and social contexts, perhaps what is needed are exercises that cultivate attention. To attend to the world, or to cultivate one's capacity for attention, is to be able to begin to see a sense of complexity and mystery that exceeds our habituated modes of receiving and responding. As Simone Weil describes, attention "consists of suspending our thought, leaving it detached, empty, and ready to be penetrated by the object."[20] She proposes that if our intellectual efforts are grounded in joy, desire, and love—and not simply the strain of the will or the expectation of accomplishment—then school studies can cultivate this capability and be harnessed for prayer. The transformation resulting from the activities of attention extends from one's relationship to the divine to the ability to recognize the plight of others and to be with them in their suffering.

Attention can inform and be formed by every aspect of our lives. Janet Martin Soskice suggests that in place of traditionally drawn divisions between active and contemplative work, one could easily approach caring for children, for example, as a practice that cultivates attention. Soskice directs our focus away from the simply metaphysical to the epiphanies of the ordinary as she writes that "the object of attention is not a changeless truth so much as a moving target."[21] Fishbane envisions that "theological attentiveness will thus be geared to the minutiae of our lives—all the time. In this manner, we may confess at each moment: 'Here I am,' living thankfully among God's gifts."[22] In this way theological practice is integral to an entire way of living, and it is not a specialized and compartmentalized form of cognitive activity or reflection.

Gardening, on Wirzba's account, is a practice that both demands and cultivates such attention. Though one might change the landscape through gardening, it cannot be done in a way that disregards the constraints of place or the singularity of a plant. Tending to a garden "requires attention and patience and a tremendous amount of detailed knowledge about soil and plant and animal life, not to mention weather and the peculiarities of different growing zones and topographies."[23] It is impossible to control the flourishing of a plant; it is only possible to aid, to respond to its needs, to work *with* it.[24] The pedagogy of the garden is quite different from that of the monoculture of the yard. Michael Pollan observes, "For if lawn mowing feels like copying the same sentence over and over, gardening is like writing out new ones, an infinitely variable process of invention and discovery."[25]

This practice has deep roots in our theological and philosophical imaginary. Epicurus located his school in a garden and demanded that his disciples work in it. In so doing, the disciples were not simply educated about their capacity for creation, but also about the limits of their creaturely life—as they learned of the connections between their labor and the nurture of their friends and the mortality of all matter.[26] Likewise, Jesus often gathered with his disciples in a garden and spoke of himself as a vine and his disciples as branches in need of pruning (John 18:12; 15:1–11). In the Yahwist account of Genesis, the vocation of humanity is to till and keep the garden (Gen. 2:15). Trimming back some of the overgrowth of time, Wirzba notes that the Hebrew word for "till" in this verse also could be translated as "serve."[27] This service, however, is not a matter of simple obedience, but is more a matter of attentive relationship, of care.

The activity of gardening can serve to cultivate alternative capabilities that emerge out of a changed relationship with creation and with others. Whereas the lawn seems to display a simultaneous arrogance (by disregarding location) and thoughtlessness (by anxiously conforming), the garden demands that its student become humble and attentive. While this work might frustrate our desires for control, order, and immediacy, it might also open spaces of wonder, mystery, and awe.[28] Toiling with our hands and responding to the limits of creation might be one way to begin to reclaim our creaturely creativity without destroying the world in which we live.

The Grass of the Field

The American lawn might serve as both school and symbol of a way of life, of forms of cultivation, of a culture whose arrogant assertion of its own power covers up its anxiety about disorder and stunts its ability to care for itself and its world. This botanical formation reveals the reversal of the prophetic call of Second Isaiah, which declares, "all people are grass, their constancy is like the flower of the field. The grass withers, the flower fades . . . but the word of our God will stand forever" (Isa. 40:7b–8). The American lawn does not flower, and great efforts are made to see that it does not fade. In contrast to the long grasses of a prairie or the flowering grasses of a meadow, the ideal yard is kept short, uniform, and evergreen.

James connects this image of the fleeting "flower of the field" to the lives of the rich (James 1:9). He writes, "It is the same way with the rich; in the midst of a busy life they will wither away" (1:11). Their practices and outlook wither because they fail both to see the manner in which everything is a gift from God and "to care for orphans and widows" in their affliction (1:17, 1:27). The undying condition of the American lawn betrays the hope that it is not the word of God, but the people and their social order that live forever. Encompassing our communities, stretching along the highways, and exported through the imaginary of our media, the American lawn is the expression and catechesis of a realized eschatology that identifies the Kingdom of God with the Manifest Destiny of consumerism.

While gardens can provide spaces of alternative instruction, if part of what is at stake is an underlying inability to deal with disorder, difference, and strangeness, then such forms of care might not be entirely adequate. Gardens typically require a predetermined plan and a great deal of initial uprooting and ongoing weeding. Weil warns that one of the great obstacles to attention is our tendency toward being "too active."[29] While the activity of gardening does require a great deal of responsiveness and humility, perhaps it might still need to be complemented by spaces that further abdicate our creaturely creativity. As Wirzba notes, developing attention is about belonging to a certain place; it is about living into one's membership.[30] If this space always follows the pattern of the yard or the garden, then the limitations of membership might be drawn in too rigid or predetermined a way.

In the Sermon on the Mount, Christ counsels that we turn our attention to a different habitat: "Consider the lilies of the field, how they grow; they neither toil nor spin, yet I tell you, even Solomon in all his glory was not clothed like one of these. But if God so clothes the grass of the field, which is alive today and tomorrow is thrown in the oven, will he not much more clothe you?" (Matt. 6:28–30). The withering of the grass and flowers of the field is accompanied by the charge to learn from the way that they flourish. Christ does not simply condemn the transitory character of our lives and our flesh, but points toward a different way of attending to that which is fleeting. The lilies of the field do not simply toil, but they glean from the bounty of the wild.

Søren Kierkegaard emphasizes the relevance of these counsels for daily life by proposing that turning one's *attention* to the lilies of the field and the birds of the air might serve to transform the busy, destructive, and anxious character of modern life. Though much of our quotidian experience is oriented around different modes of comparison—in which we evaluate our merit or worth in terms of the roles that we play or the status we possess—Kierkegaard proposes that by turning to the lilies of the field we might receive a different education. He writes that "if the person in distress actually gives his attention to the lilies and the birds and their life and forgets himself in contemplation of their life," he learns something about himself.[31] He learns that while the lilies do not toil—they do not busy them-

selves with needless and arrogant accumulation and comparison—
they are beautiful and glorious. Similarly, a human being is not glorious because of its own creations, but because of its creatureliness;
it is something that is created by God and that is in a vulnerable and
close relationship with creation.

In order to learn from the flowers and grass of the field, it might be
necessary to actually be around fields, to inhabit spaces in which the
ostensible chaos of the wild can sprout, bloom, and instruct those
who are attentive to it. Perhaps the grass of the field could serve to
cultivate an attention that does not simply understand or discern
proper order, but that sees the beauty in strangeness. The grass of the
field might serve as a space where we can begin to displace "the classical sense of an ordered and symmetrical beauty" with the beauty
of the strange.[32] As Karmen MacKendrick proposes, this is a beauty
that stops us and unsettles our habituated attitudes; it "makes us re-
arrange our perceptions, makes us see again."[33] This is a matter "of
looking again at what was boringly familiar" and of an "attentive
awakening" through acts of hospitality.[34] As we attend to the grass
and lilies of the field, they welcome us back into the strangeness of
our selves, as we do not possess or order, but only receive ourselves.
This in turn helps us to welcome the strangeness of others and those
who are strangers to our order and memberships. The care of the self
and the care of the soil extend to the care of others.

Such spaces need not be opposed to gardens. As Wirzba has argued,
the veneration of the wild has often led to a hardening of the line between nature and culture and therefore to an increased distance between our daily lives and attention to creation.[35] What is needed is a
diversity of spaces and porous boundaries between them.[36] In place
of an unsustainable homogeneous habitat that demands inattentive
care, we need a plurality of spaces in which we can educate capabilities and cultivate relationships. While some of these spaces require
the careful activity of gardening, other fields simply demand our attention. Neither of these requires the renunciation of our creativity,
but they do demand the embrace of our fundamental relatedness, of
our place as creatures in creation.

So much hangs on Keller's question: "Can we inhabit this liminal space of discourse, articulating an edge that does not so much cut as *fold*, complicating rather than separating our divergent times and spaces?"[37] We have been habituated to be uneasy with this kind of complexity and disorder. But perhaps it is time that we challenge the exercises and built environments that have trained us to think this way. In order to face the current ecological and existential crises it is vital not only that we inhabit this space of discourse, but that our imaginative and creative discourse address and respond to the places in which we live and act. Therefore, philosophical theology might begin by thinking and acting in a variety of spaces with attention and care. While this does not mean that we should distrust the positive associations that we might have with the smell of freshly cut grass, it does seem important to begin to look again at the habitats around us that are so often made invisible by their nearness. In attending to these places, we might ask: what exactly is being cultivated and worshiped in this space, and what is being excluded and sacrificed?

Dream Writing Beyond a Wounded World: Topographies of the Eco-Divine

Susan Pyke

If it is accepted that pro-environmental actions are related to positive perceptions of the external environment, then it can be argued that literature has the potential to shift readers' preparedness to act for constructive ecological change.[1] Texts that discern communications from nonhuman matter play a particularly interesting role in this conditional context. Anne Carson's passionate contemporary poem "The Glass Essay," together with Emily Brontë's excessive nineteenth-century novel *Wuthering Heights*, forms a case in point.[2] Carson, like Brontë, describes the effect of an exchange between the human and nonhuman, which necessarily occurs beyond the limits of signification, and her poem's engagement with Brontë's novel suggests in itself that artwork can move readers to action. The productivity in Carson's rhizome-like iteration, where she reads *Wuthering Heights* and responds by writing "The Glass Essay," suggests the possibility for further actions as others read her revision or, more precisely, re-vision.

By accepting the sentience of the nonhuman, the writing of Carson and Brontë goes beyond the restrictive exclusivity of Father God, fissuring anthropo-theological ideas of human dominion over the earth and gesturing toward an eco-divine. "Divine" is used here in allegiance with Luce Irigaray's noncontained "becoming divine"[3] because of the way Carson's poem, with Brontë's *Wuthering Heights*, troubles the prevailing dualisms present in Judeo-Christian thinking. Irigaray's vision is of an always-becoming divine, uncontained by words, reaching toward a material yet unutterable connection with the divine. She writes against the male centricity of theology, but her vision does have sympathies with Christian theologies that counter-think the centrality of the human.[4] My question is whether Carson's writing of the eco-divine might move readers toward a greater pro-environmental consciousness, just as her reading of *Wuthering Heights* seems to influence the pro-environmental consciousness of "The Glass Essay." If this is the case, eco-divine literature might lead to more sustainable lifeways by encouraging a closer attentiveness to country.[5]

Over the course of Carson's poem, her speaker leaves behind the damage of a restrictive love affair through attentive exchanges with the fragile, brilliant, and communicative Canadian moor. She experiences a coaffective flow between habitation and self, an "emplacement,"[6] and this generative communicative process is facilitated through the speaker's engagement with Brontë's *Wuthering Heights*. Her reparation begins at the point where she starts "turning into Emily Brontë." This shape-shifting potential increasingly empowers her to "become," beyond the limitations of her contained self.

The productive element for my argument is the way Brontë and Carson both depict their central protagonist as part of a community of matter that includes flora, fauna, land, sea, and air.[7] This post-human "country" approaches a proper noun;[8] it is not the kind of country that exists through politically constructed borders. Thinking "country," rather than place, habitat, or environment, venerates a related usage of country integral to the culture of many Aboriginal and Torres Strait Islanders. However, while both Carson's speaker and Brontë's central protagonist, Cathy, participate in the matrices of interdependent relationships, "country" approaches a proper noun.[9] I do not suggest their participation approaches the deeply unique relationships available to past, present, and emerging Aboriginal and Torres Strait Islander communities. Less intrusively, my respect for the potentiality expressed in such relationships with country, where human listening is privileged over human speaking, simply gives me faith in the possibility of an always-becoming eco-divine, a divine of change, that listens to the echo (eco) of country that communicates the more (moor) of becoming. The relationships Carson's speaker and Brontë's Cathy have with country do not assimilate them into preexisting relationships outside their ancestry, but they do become attuned to the specificities of their own personal emplacement and their own ways of being as they listen to country speaking to them in its own way. This affect goes beyond ventriloquism. Carson and Brontë do not speak *for* country. John Ruskin's description of the pathetic fallacy and metaphor has a different meaning in this context.[10] His concepts speak in human terms, where the voice of country speaks to what humans cannot signify.

So it is that there are no detailed descriptions of the communications of country in either the novel or the poem. Carson continues Brontë's acceptance of the impossibility of writing country as it "is," accepting that its complexity goes beyond human understanding.[11] They offer country's communicative exchanges as *felt*, depicting human affect, rather than nonhuman communications, to demonstrate the possibilities in moving toward a part of country that has meaning for individual subjectivities.

Kate Rigby's literary ecology acknowledges that texts can think country in this way through her suggestion that country "talks back."[12] Rigby's analysis of nineteenth-century German literature shows that just as texts influence interactions with country, so too does country influence interactions with texts. Thinking about country "talking back" infers backchat, and if backchat is about questioning the assumptions of the dominant voice, listening to the backchat of country makes sense. After all, the dominant voice of the human has proven unable to do justice to country, including its human component. There is little need to further gloss the fact that human fossil-fuel consumption, deforestation, and overpopulation are making the earth less habitable for the human species, as well as for a vast majority of the other species making up this world's earth.

Humans are part of a layer in the making called, with hubris and/
or shame, the Anthropocene. Taking this seriously means taking it
personally, and the same goes for country's backchat. It is a particular, not a general, response.

In Brontë's *Wuthering Heights* and Carson's "The Glass Essay" a
counter-voice, the voice of country, emerges through specific communicative mutual encounters, illustrating Rigby's argument that
topographies, or the writings of place, include the possibility of
country speaking "in its own voice and in its own regionally specific idiom."[13] In *Wuthering Heights* these encounters are represented through the responses of Cathy, who listens to the flow of the
moor and gives it the flow of her life. Carson's poem also communicates encounters with country through affect. Readers understand
the moor assists the speaker's evolution as she sheds the civilized
comfort of ego-contained flesh to move beyond the limitations of
her gaze-tortured skin. Because "The Glass Essay" explicitly references *Wuthering Heights*, a doubling effect runs between the love-
lorn Cathy and the love torn speaker. The communications of the
Canadian moor change the way her speaker thinks, and this is me-
diated through Brontë's depiction of the Yorkshire moor's affect on
Cathy.

Carson's speaker acknowledges the relationship between Cathy
and the country of her childhood, primarily by evoking their names,
inviting readers of the poem to bring their own readings of *Wuthering Heights* into "The Glass Essay." To my reading, this relationship
evidences most strongly in the section of the book where Cathy is
close to childbirth and equally close to death, bound to her new
home of Thrushcross Grange and yearning for a reunion with her
deeply loved Heathcliff and the moor that was part of their earlier
being. She listens to "the full, mellow flow of the beck in the valley" that "always sounded on quiet days following a great thaw or
a season of steady rain."[14] The sounding of the beck, its flowing discourse, calls Cathy to respond by changing her relationship to her
body. The flow of the beck is more than a metaphor for Cathy's
inner turmoil. Its course helps Cathy fight against the threat of a

motherhood lived through the confines of the controlling Edgar. She searches for a route of return that will take her to her childhood contentedness, with the headstrong resistance of breeding salmon, and finds her way there through dreams and self-starvation. Her desire is to leave her body's "shattered prison" and become "always" part of the "glorious world" rather than "yearning for it through the walls of an aching heart."[15] She chooses to return, in the only way open to her, to being part of the moor.

Brontë scholar Enid Duthie describes the beck as being, for Cathy, "the nearest approach to peace" known to her "restless" nature.[16] There is an important caveat in Duthie's description: near to peace, not peaceful. Duthie's hesitation points to the dangerous undercurrents that create the "murmur of the beck . . . its ripples and its gurgling."[17] In Yorkshire a beck can shift from a rivulet to a torrent in a few short hours, and its waters have been strong and lasting enough to carve clefts in the rise of the hills. Similarly, water directs human steps and forms the flesh of human bodies, and, specifically, it directs Cathy to find a way out of her constraints so she might find a less hurtful shape.

Brontë pointedly contrasts Cathy's interrelationship with country to the insentience of the novel's first narrator. The incapacity of the aptly named Lockwood, a body of fixed matter, to listen to country is made evident from the first movement of the book. His "tacit insult" of making faces at a "ruffianly bitch" is an attempt at mastery that ends with the nursing dog breaking into a "fury" and leaping up at him. Lockwood dodges behind the table to avoid this backchat and parries with a poker as "half-a-dozen four-footed fiends" set upon his coat tail and heels.[18] Only his pride is hurt; his self-centered humanity remains. Lockwood's antipathy to the country Cathy pines for extends well past Heathcliff's "vigilant" dogs. His exposure to the Yorkshire moor fills him with dismay because the only voice available to him comes from "bleak winds and bitter northern skies." He bitterly misses "human physiognomy,"[19] which talks in a language he is prepared to hear. Unlike Cathy, who keenly listens to the moor in all seasons, Lockwood is deafened by the self-limited body of a reasonable, Enlightened man. He is displaced, anxious to re-form his setting; he wants to recentralize his humanity.

By negatively comparing Lockwood's displacement to Cathy's emplacement, Brontë contests humankind's "God given" right to conquer the earth, as legitimized through the theology of dominion, writing against an insentient humanity that tangles and tears these tender interconnections. Similarly, the Canadian moor is in no way subservient to, or separate from, Carson's speaker. Instead, it is an active change agent, making clear communications that affect the way she understands her life. "Being with" the moors of her childhood, through and with *Wuthering Heights*, helps her find new ways to navigate through the hurtful blockages in her life.

Carson offers beautiful, if frightening, descriptions of her speaker's emotional response to the healing offered by communications from the Canadian mud, trees, ice, and air. As with Brontë's depictions of the affective Yorkshire moor, in Carson's poem there are no detailed landscape descriptions. Country is present through feelings expressed by the speaker, doubled through the productivity of intertextuality. The speaker's engagement with Brontë's novel assists her to respond to the communications of the moor. She does not move on her own; she moves with Cathy. Her escape from destructive human desire, based on coercion, which has subsumed her ability to make her own choices, begins, as life does, with the mother.

> She lives on a moor in the north.
> She lives alone.
> Spring opens like a blade there . . .
> Whenever I visit my mother
> I feel I am turning into Emily Brontë,
>
> my lonely life around me like a moor.[20]

Carson blurs the speaker, the speaker's mother, Brontë, and the cut of spring that opens the speaker to a liberating darkness, providing something of an answer here to Luce Irigaray's dare to cut through all the "layers" that create selfhood. Irigaray warns of the risk involved in such rebellion. Beyond the layers waits "a night far deeper than your day can imagine," a night of possibilities that weave "the fabric of the world, its cycles and revolutions" into different always-becoming potentialities.[21] As is made clear in the poem's final lines, there is no guarantee that moving toward such change leads to a predictable or benign eco-divine. All that is promised is a shift away from the limits of Father God domination and its corollary, human dominion.

As the poem's narrative unfolds, the speaker's affinity for the moor takes her to a post-human love beyond the cruelty of love focused on the gratification of the male ego.[22] Like Cathy, and unlike Lockwood, Carson's speaker opens to a mutuality between self and country, and this occurs through visions nurtured through country and text. Through this meditative and embodied work she is led to a number of humbling yet liberating epiphanies. Most strikingly, as she participates in her final dreamscape, she finds herself flailed free of the flesh that limits her to the male gaze. Her bones

> stood forth silver and necessary.
> It was not my body, not a woman's body, it was the body of
>   us all.
> It walked out of the light.[23]

Carson's words resonate with David Abram's "earthen silence of shadows and bones and leaf."[24] Her "body of us all" moves beyond mind/body dualisms that privilege the "soul" or "spirit" over the "body" or the "earth," and in this way, "The Glass Essay" refutes more limited readings of Brontë's novel as a struggle between good and evil. Carson's poem underlines the intent uncertainty of *Wuthering Heights*, continuing Brontë's project of stepping beyond the reasonable rulings of the Enlightenment. There are no binaries to divide Brontë's Cathy from Heathcliff, Edgar, or the moor, nor is Carson's speaker separated from her ex-lover Law, or the moor, or, indeed, her mother. Instead, these texts open to the abyss of self-

dissolution, an abyss that forms part of the productive w/hole Irigaray describes. The darkness holds the crypt of the mother,[25] and the
possibilities of this crypt infiltrate the blinding enclosure of a Father
God formed only of light. The hole of the abyss is part of the whole,
just as organic matter depends on the dark matter of the earth to
thrive, as well as the light of day. Opening the self to the w/hole
crypt allows barely perceptible assemblages to take the self beyond
the limits of the singleton required in the concept of humanity.[26] By
doing away with the attachment to a specific body's skin, by seeking interdependence, not independence, the entrapment of self can
dissolve. Such dissolution is not limited to human selves. Like the
ghostly wind that carries Cathy's voice to Heathcliff as he digs his
way into her grave, the body is part of an "all" beyond the limits of
the "self"-ish human. Brontë moves in this direction with Cathy's
famous declaration, "I *am* Heathcliff."[27] They are one and the same
just as she *is* the moor.

   This coaffective move requires a mutual or shared response.[28]
The way Carson's raw poem "gets" the acceptance of the sentience
of country in *Wuthering Heights* demonstrates what psychoanalytic
theorist Bracha Ettinger refers to as a "conductable affectivity." For
Ettinger this affectivity occurs with the emergence of the unutterable, in a shared space of "unthought subknowledge."[29] If Ettinger's
thinking is extended to include encounters with the nonhuman,
encounters that also cannot be spoken, then the affect Cathy and
Carson's speaker experience, beyond human signification, has this
kind of affective conductivity. By keeping company with *Wuthering
Heights*, the speaker shares Cathy's permeation into the moor. When
Carson's speaker starts "turning into Emily Brontë," she also begins
"turning into" the moor, then her mother, and (never quite) finally,
the dynamic and always becoming "body of us all."[30] The intertextual strength of this w/hole of country in *Wuthering Heights* and
"The Glass Essay" creates a shared sensible space that approaches
that which Ettinger calls the matrixial borderspace.

   The matrixial relations between these two works and country
can be further understood through Hélène Cixous's application of a

poetics of literary theory to her favored "dream writers," who, like Brontë and Carson, offer "a body-to-body" transport to the country of dreams through affect.[31] Lucinda Miller, in her punchy preface to a recent reprint of *Wuthering Heights*, refers to Cathy's description of the dreams that stay with her, the ones that go through her body like wine into water, then defines *Wuthering Heights* as "just that sort of dream."[32] Indeed, like two separate but related dreams, Brontë's novel and Carson's poem have unexpected, specific perspectives that can be recognized, even while their localized materiality is their own. They offer a topographical invocation to the body,[33] through the dream-traveling mind, and readers' acceptance of the sentience of country they depict will be contingent on previous experiences with country and text. I read how country moves Brontë's Cathy, how her heart aches to be part of the unfettered moor, how Heathcliff hears her sigh as part of the wind, and I share this affect in ways grounded in the materiality of my own interchanges with country and text. I ache with the changes to my rocky childhood country and listen to what the wind tells of its rolled-flat state. This affect is strengthened when I read Cathy's embodied response through the affect it has on Carson's speaker. Through these texts I become closer to the matter of the country I am most part of; it is a doubled response, threaded with Carson's speaker's response to Cathy's response. Carson's Canadian moor is doubled through Brontë's Yorkshire moor, then doubled again through the country that gives me backchat. I shift in the company of both texts, they awaken me, and I find my body moved to a different place. This intertextual shared dream writing of country grounds me in materiality, and, at the same time, it expands my understanding of immanence beyond the skin of my words.

It is no easy shift, going from words to affect, harder still to trace what might move affect to action. I suggest nothing more tangible than possibility. However, Cixous's idea of dream-writing does give this possibility an ethical imperative. If the dream-writing in *Wuthering Heights* and "The Glass Essay" shatters my slumber, and this awakening cuts through the layers that keep me from the crypt-like night of possibility, then, as dream analyst Carl Jung has argued, I am obliged to take action.[34] It is the remembered dream that matters. Dream writing of the eco-divine that awakens me to a new awareness of my mutuality with country requires me, ethically, to live differently with country.[35]

There is a cross here, from literary theory to eco-philosophy, for, as Freya Mathews argues, once humans open to the "flow" of the communicative exchange, through the embodied response,[36] they become more aware of the grace in country. The more I respond to this grace with action, the more I become emplaced. This flow is invited in through positive inaction, where, like Cathy and Carson's speaker, I remain still and open, honed by the blade of the seasons. This might, I hope, lead me to productive shifts in my lifeways.[37] Intertextual works like "The Glass Essay" help orient such readers toward earthly generation, rather than degeneration, and each fresh invocation broadens the grace of flow. Once the sentience of country is welcomed, rather than refused, acting against the interest of country becomes an act of betrayal, both of the self and beyond the self, in the "all" or the w/hole of country.

The beautiful fury of Carson's Brontë-inspired dream writing shows how this flow, strengthened through the literary response, might help bodies move toward lifeways that take the more-than-human into account. Carson's re-visioning of Brontë's text suggests an always-becoming eco-divine that encourages readers to hear country differently. Importantly, there is no hierarchy in productive intertexuality. *Wuthering Heights* is not limited to root-bound re-productions that graft onto day-lit branches.[38] "The Glass Essay" responds to Brontë's novel's dark rhizome-like openness, and it also creates its own growth. Both texts playfully create new sites of mutuality beyond the limits of human signification; both engage at the roots of a psyche that is not contained to the self, and they do this in singular ways.

My own novel, *The Dead Country*, enters this multiplicity by playfully subverting the plot of *Wuthering Heights* into an imaginative exploration of the sentient country I know beyond utterance. My acceptance of more-than-human sentience is made apparent when the primary protagonist, Bai Hua, flees from a ghost, felt within herself and by her landlord, Ketty. She shifts from blind panic when she becomes aware of her surroundings.

> I hear crying under the road. The trees are pushing to me, telling me to listen to the dirt, down deep. They want me to open to the ghosts inside this dirt; they want me to do my ghost listening.
> I can do this but I don't want to do this. The trees are too close, making big noise, my head can't hear nothing but this asking. No! None of my business, okay?

Despite her fear, obfuscated as disdain, Bai Hua does, in the end, listen, and the backchatting silence of country has its way with her:

> the white grass is too quiet, no juice inside, and crispy fried brown ferns are telling nothing taking nowhere. The fat old trees with the big white bodies and little grey-green leaves not making one sound, telling me listen from how they stand still.
> Okay, I'm listening, stupid dirt.

As Bai Hua tunes to the flow of her new habitat, she begins to learn more of her potential to live differently where she is situated. Bai Hua's growth suggests what can be gained from staying still, even when running seems the best thing to do.

I know my literary response to Brontë's eco-divine text is a very small card to play in the end-game humans are playing with the earth's commons. My suggestion is simply this: if eco-divine texts help readers make the time to listen to the country that backchats to them, then that which is eco-divine in country might, with grace, move readers from words to affect, and so to actions that go beyond the limits of what might be said. As I continue to read and to write, I extend the same hope to that part of country that I still insist on calling myself.

Notes

Introduction: The Human Place in the Natural World
Brian Treanor

1. Gary Snyder, *The Gary Snyder Reader: Prose, Poetry, and Translations* (Berkeley, Calif.: Counterpoint, 2000), 169.

2. Wallace Stegner, *Angle of Repose* (New York: Penguin, 1971), 159.

3. St. Irenaeus, *The Writings of Irenaeus*, ed. Alexander Roberts and James Donaldson (Berkeley, Calif.: Apocraphyle Press, 2007).

4. St. Augustine, *Confessions*, trans. Henry Chadwick (London: Oxford, 1992); see books IV and VI.

5. Thomas Aquinas, *Summa Theologica*, Prima Pars, 93, 1–9.

6. Karl Barth, *Church Dogmatics III/2* (Edinburgh: T. and T. Clark, 1956–60).

7. Lynn White, "The Historical Roots of our Ecologic Crisis," *Science* 10 (March 1967): 1203–7.

8. White, "Historical Roots," 1205.

9. René Descartes, *Meditations on First Philosophy*, trans. Donald Cress (Indianapolis: Hackett, 1993) 97 (as only one representative example).

10. Plato, *Phaedo*, trans. G. M. A. Grube (Indianapolis: Hackett, 1981), 98, 100–2.

11. For more on this subject, see David Clough, *On Animals*, vol.1, *Systematic Theology* (London: T. and T. Clark, 2012).

12. See Percy Bysshe Shelley, "Ozymandias," in *The Norton Anthology of Poetry* (New York: Norton, 2004).

13. This turn of phrase is borrowed from Jack Caputo, who uses it while reflecting on Nietzsche: "Once upon a time, in some out of the way corner of that universe which is dispersed into numberless twinkling solar systems, there was a star upon which clever beasts invented knowing. That was the most arrogant and mendacious minute of 'world history,' but nevertheless it was only a minute. After nature had drawn a few breaths, the star cooled and congealed, and the clever beasts had to die"; *Philosophy and Truth: Selections from Nietzsche's Notebooks of the Early 1870s*, ed. and trans. Daniel Breazeale (Atlantic Highlands, N.J.: Humanities Press, 1979), 79. Caputo reflects on the significance of this passage in John D. Caputo, *Against Ethics* (Bloomington: Indiana University Press, 1993), 16–17.

14. Erzaim Kohák writes, "Both in principle and as a matter of historical fact, alienation sets in when humans lose their awareness of the presence of God and persuade themselves to view the cosmos no longer as a creation, endowed with value in the order of being, a purpose in the order of time and a moral sense in the order of eternity, but as a cosmic accident, meaningless and mechanical. Then nature comes to appear as absurd and we ourselves as futile within it"; Kohák, *The Embers and the Stars* (Chicago: University of Chicago Press, 1984), 183. For an extended philosophical treatment of the disappearance of creation in modern philosophy, see Louis Dupré's *Passage to Modernity: An Essay in the Hermeneutics of Nature and Culture* (New Haven: Yale University Press, 1993).

15. Charles Darwin, *The Descent of Man, and Selection in Relation to Sex* (Princeton: Princeton University Press, 1981), 405.

16. Martin Heidegger, *Being and Time*, trans. John Macquarrie and Edward Robinson (San Francisco: Harper San Francisco, 1962). Heidegger also argues that humans have a "world" (*welt*), but animals are "poor in world," that is, they do not have a world in the sense that humans do. Here again, the being of humans is sharply differentiated from the being of animals: while humans are creative and world-forming, animals are poor in world (and inert matter would be entirely worldless); see Matthew Calarco, *Zoographies: The Question of the Animal from Heidegger to Derrida* (New York: Columbia University Press, 2008).

17. Heidegger, *Poetry, Language, Thought*, trans. Albert Hofstadter (New York: Harper, 2001).

18. See Jacques Derrida, *Archive Fever: A Freudian Impression*, trans. Eric Prenowitz (Chicago: University of Chicago Press, 1995).

19. Snyder, *The Practice of the Wild* (Berkeley, Calif.: Counterpoint, 1990), 31.

20. Edward S. Casey, *Getting Back Into Place* (Bloomington: Indiana University Press, 1993).

21. Casey, *Getting Back Into Place*, 13. Casey goes so far as to make place co-primordial with God.

22. See, for example, Forrest Clingerman, "Interpreting Heaven and Earth: The Theological Construction of Nature, Place, and the Built Environment," in *Nature, Space, and the Sacred: Transdisciplinary Perspectives*, ed. S. Bergman, P. M. Scott, M. Jansdotter Samuelson, and H. Bedford-Strohm (London: Ashgate, 2009).

23. Clingerman, "Interpreting Heaven and Earth," 47.

24. Ben Okri, in an interview in the *Independent*, March 23, 1993.

25. See Robert Mugerauer, *Heidegger and Homecoming: The Leitmotif in the Later Writings* (Toronto: University of Toronto Press, 2008); also see Richard Kearney, *Anatheism* (New York: Columbia University Press, 2010), 13.

26. Kohák, *The Embers and the Stars* (Chicago: University of Chicago Press, 1984), 84.

Creation, Creativity, and Creatureliness: The Wisdom of Finite Existence
Rowan Williams

1. Also translated as *The Pillar and the Ground of Truth*, trans. Boris Jackim (Princeton: Princeton University Press, 2004).

2. See Athanasius, *The Orations of Athanasius Against the Arians* (Cambridge: Cambridge University Press, 2014).

3. Sergei Bulgakov, *The Lamb of God*, trans. Boris Jackim (Grand Rapids: Eerdmans, 2008).

4. Bulgakov, *Philosophy of Economy: The World as Household*, trans. Catherine Evtuhov (New Haven: Yale University Press, 2000).

Rowan Williams and Ecological Rationality
Jarrod Longbons

1. Slavoj Žižek, "Joe Public v. the Volcano," *New Statesman*, April 29, 2010, accessed June 1, 2011, http://www.newstatesman.com/

environment/2010/05/essay-nature-catastrophe. The time when humans are such geographical agents is called the "anthropocene" era.

2. Žižek, "Joe Public v. the Volcano."

3. For Žižek's account of "the environment as a series of catastrophes," see his reflection on oil in "Censorship Today: Violence or Ecology as a New Opium for the Masses," accessed March 19, 2012, http://www .lacan.com/zizecology1.htm.

4. Žižek, "Censorship Today: Violence or Ecology a New Opium for the Masses."

5. Žižek, *Living in the End Times* (London: Verso, 2011), 80.

6. Aldo Leopold, *A Sand County Almanac* (New York: Oxford University Press, 1949), 224–25.

7. Hans Jonas, *The Phenomenon of Life: Toward a Philosophical Biology* (Evanston, Ill.: Northwestern University Press, 2001), 13.

8. See Bruno Latour, *We Have Never Been Modern* (Cambridge, Mass.: Harvard University Press, 1993).

9. René Descartes, *Discourse on Method* (Miami: B. N. Publishing, 2007), 50.

10. See Gísli Pálsson, "Human–Environment Relations: Orientalism, Paternalism, and Communalism," in *Nature and Society: Anthropological Perspectives*, ed. Philippe Descola and Gísli Pálsson (London: Routledge, 1996), 68.

11. Descola and Pálsson, *Nature and Society*, 3.

12. See Žižek's portion of the documentary *Examined Life*, directed by Astra Taylor (2008; Canada: Sphinx Productions), DVD. True, humans are "natural" to earth's biosphere, and with humanity comes human making. But this does not mean that all human making is harmonious with "nature."

13. This is a classical distinction between nature and art. The nature of a tree is not to be a bed; only when acted upon, artistically, can a tree be made into a bed.

14. Human technology is most harmonious with nature when it is in accordance with the natural rhythms of the world; see Erazim Kohák, *The Embers and Stars* (Chicago: University of Chicago Press, 1984). Kohak discusses the way people light houses at night. The flick of an electrical switch completely eradicates the darkness. It gives the impression that it is not nighttime. Lanterns, on the other hand, light only a portion of

a room. When lit, darkness falls around the power of its glow. It is still clear that it is the rhythm of nighttime. This small example is extended to larger issues. For Kohák, the issue is not that humans should shun technology, for even fire is technology, but rather that humans must ask if their technology goes harmoniously with nature or against it.

15. See Sergei Bulgakov, "Heroism and the Spiritual Struggle," in *Sergei Bulgakov: Towards a Russian Political Theology*, ed. Rowan Williams (London: T. and T. Clark, 1999), 81–82.

16. To illustrate a parasitic view of humans and their work, consider Agent Smith's speech to Morpheus in *The Matrix*: "I'd like to share a revelation I've had during my time here. It came to me when I tried to classify your species. I realized that you're not actually mammals. Every mammal on this planet instinctively develops a natural equilibrium with the surrounding environment, but you humans do not. You move to an area and you multiply and multiply until every natural resource is consumed. The only way you can survive is to spread to another area. There is another organism on this planet that follows the same pattern. Do you know what it is? A virus. Human beings are a disease, a cancer of this planet. You are a plague, and we are the cure"; *Matrix*, directed by Andy Wachowski and Larry Wachowski (1999; United States: Warner Bros. Pictures), DVD.

17. See Wendell Berry, *The Art of the Commonplace: The Agrarian Essays*, ed. Norman Wirzba (Berkeley, Calif.: Counterpoint, 2002), 279–92.

18. Ani DiFranco identifies more with those in the animal world "who only take what they need"; Ani DiFranco, "Animal," *Educated Guess*, 2004, Righteous Babe, CD.

19. Rowan Williams, "Renewing the Face of the Earth: Human Responsibility and the Environment," The Ebor Lecture, York Minster, posted at http://www.archbishopofcanterbury.org/articles.php/816/renewing-the -face-of-the-earth-human-responsibility-and-the-environment on Wednesday March 25, 2009, accessed June 1, 2011.

20. "Outflanks" is John Milbank's terminology for overcoming this dualism by the theological doctrine of creation; Lecture 8 of the 2011 Stanton Lectures, Cambridge, posted at http://theologyphilosophycentre.co.uk/ papers/Milbank_StantonLecture8.pdf, accessed July 23, 2012.

21. Namely, various dispensational theologies that say the world will end in destruction and Christians will escape this demise as they go on to heaven; see Michael S. Northcott's chapter "The Dominion Lie: How

Millennial Theology Erodes Creation Care," in *Diversity and Dominion: Dialogues in Ecology, Ethics and Theology*, ed. Kyle S. Van Houtan and Michael S. Northcott (Eugene, Ore.: Wipf and Stock, 2010), 89–108.

22. The land needs to be understood as more than just earth. The Bible, like Leopold's "Land Ethic," includes all nonhuman creatures in the term *land*; it means the entire biotic community.

23. Leviticus 26:34–35, NIV.

24. Williams, "Renewing the Face of the Earth."

25. Rachel Wells, "Attack of the Flying Fish," *Illinois Times*, June 24, 2010, accessed June 25, 2010, http://www.illinoistimes.com/Springfield/article-7433-attack-of-the-flying-fish.html.

26. Affecting "nature" through human activity is what Bill McKibben calls "the end of nature" in his work *The End of Nature* (New York: Random House, 2006), because once humans can cause this type of change (though always powerless to control it), nature is no longer constant and indifferent to human life.

27. David Burrell, "Analogy, Creation, and Theological Language," in *The Theology of Thomas Aquinas*, ed. Rik Van Nieuwenhove and Joseph Wawrykow (Notre Dame, Ind.: University of Notre Dame Press, 2005), 87.

28. Williams, "Renewing the Face of the Earth." This is true even with the Christian conviction that humans are fulfillment of the creation of the universe: beings who are to perfect creation itself.

29. Ibid.

30. Alexander Schmemann, *For the Life of the World* (New York: St. Vladimir's Seminary Press, 1997).

31. "Pope Benedict's environmentalism concerned with the 'ecology of man,'" *Vatican Observer*, December 20, 2009, accessed July 23, 2012, http://www.catholicnewsagency.com/news/pope_benedicts_environmentalism_concerned_with_the_ecology_of_man_vatican_observer_writes/.

32. Pope Benedict XVI, *Caritas in Veritate*, Libreria Editrice Vaticana, 2009, accessed July 19, 2012, http://www.vatican.va/holy_father/benedict_xvi/encyclicals/documents/hf_ben-xvi_enc_20090629_caritas-in-veritate_en.html.

33. Bulgakov, "The Unfading Light," in *Sergei Bulgakov: Towards a Russian Political Theology*, 138.

34. Thomas Aquinas, *Summa Theologica*, Q. 12, accessed July 23, 2012, http://www.newadvent.org/summa/1012.htm.

35. Williams, "Renewing the Face of the Earth."

36. Ibid.

37. Take for example Glenn Beck, accessed July 1, 2012, http://www .glennbeck.com/content/articles/article/198/6470/. Here, he rants about his dislike for the "environment" in a very Žižekian way. Still other commentators go further, like Rush Limbaugh, who puts forward the idea that the earth is "big enough" to handle anything that humans will do to it; see, accessed July 1, 2012, http://www.treehugger.com/files/2010/08/rush-oil -spill-solved.php; here he says that "mother nature, i.e., God and the ocean" is big enough to handle a small amount of oil spilled. Of prime importance for both commentators is human life, jobs, and government spending.

38. Ellen F. Davis, *Scripture, Culture, and Agriculture: An Agrarian Reading of the Bible* (Cambridge: Cambridge University Press, 2008), 80.

39. Berry, *Life is a Miracle: An Essay Against Modern Superstition* (Berkeley, Calif.: Counterpoint, 2000), 14.

40. Davis, *Scripture, Culture and Agriculture*. Here Davis quotes E. F. Schumacher's *Small is Beautiful: Economics as If People Mattered* (New York: HarperCollins, 1989), 146–59.

41. Bruno Latour, "An Attempt at a 'Compositionist Manifesto,'" *New Literary History* 41 (2010): 485.

42. *Jurassic Park*, directed by Steven Spielberg (1993; United States: Universal Pictures), DVD.

43. See *The Amish: A People of Preservation*, written by John L. Ruth (2000; Pennsylvania: Heritage Productions), DVD.

44. Williams, "Renewing the Face of the Earth." Williams is commenting on the work of Christos Yannaras when he uses the word "toxic" in relation to technology that detrimentally impacts the earth.

45. See Berry, *The Unsettling of America* (San Francisco: Sierra Club Books), 1996.

46. The modern connotation of "culture" signifies (but is not limited to) film, music, theater, books, education, national holidays, and food. These artifacts, among others, "cultivate" the inner life of people, while the human connection to nonhuman life rarely fits in with the modern connotation of the word.

47. Williams, "Renewing the Face of the Earth"; Williams is quoting Fulla, a Swedish entomologist from A. S. Byatt, *The Biographer's Tale* (New York: Vintage, 2001), 243–44.

48. "Mutual flourishing" for all creation is a better ecological view than "sustainability." Sustainability seeks to simply continue human activity in ways that do not destroy life. Flourishing is the aim for creation in Christian understanding, and it seeks to bring all life to greater amounts of life. This phrase is not only preferred by a Christian vision, but it promotes the health of the biosphere and truly grants dignity to the world.

49. See Norman Wirzba, *Living the Sabbath: Discovering the Rhythms of Rest and Delight* (Grand Rapids: Brazos, 2006), and Wirzba, *Food and Faith: A Theology of Eating* (Cambridge: Cambridge University Press, 2011), 43–48. In both works, Wirzba argues that the biblical social imagination of Sabbath extends beyond human communities to the nonhuman world, as well. The Sabbath imagination, he argues, is about the flourishing of all things under the creator.

50. Williams, "Renewing the Face of the Earth."

51. Ibid.

52. Dietrich Bonhoeffer calls it the "this-worldliness of Christianity"; on this he writes, "living unreservedly in life's duties, problems, successes, and failures, experiences, and perplexities. In so doing we throw ourselves completely into the arms of God, taking seriously not our own sufferings, but those of God in the world—watching with Christ in Gethsemane. That, I think, is faith; that is *metanoia*; and that is how one becomes a man and a Christian (see Jer. 45!). How can success make us arrogant, or failure lead us astray, when we share in God's suffering through a life of this kind?" Bonhoeffer, *Letters and Papers From Prison* (New York: Touchstone, 1971), 369–70.

53. Williams, *On Christian Theology* (London, Blackwell, 2000), 273.

54. Williams, "Renewing the Face of the Earth."

55. Williams, "Act of the Sake of Love" (sermon delivered at the Copenhagen Cathedral, December 13, 2009).

56. Anna Almendrala, "Toxic Tours of Los Angeles," *Huffington Post*, July 28, 2010, accessed July 20, 2012, http://www.huffingtonpost.com/2010/07/28/toxic-tours-of-los-angele_n_661289.html#s118600&title=Communities_for_a.

57. Williams, "Renewing the Face of the Earth."

58. For Žižek, ecological changes will result in changing geopolitical boundaries. He argues that we will see a time of nomadism, once again, and the only global answer for dealing with moving populations is com-

munism. His foresight, again, might be correct, but there is already a community (not an institution) that is poised to deal with global nomadism and need. This community is the universal church, which is a metanational body without borders. The church universal is poised to accept, in hospitality, people in need all around the globe.

59. See Marshall Sahlins, *The Western Illusion of Human Nature* (Cambridge: Prickly Paradigm Press, 2008).

60. Williams, "Renewing the Face of the Earth."

61. This is why St. Paul declares in Romans 8 that creation waits groaning, because of humanity's persistent denial of God.

62. Williams, "Renewing the Face of the Earth."

63. In fact, true delight leads us away from the connotation of consumption and to the reality of dining, feasting, and enjoying.

64. John of Damascus, *Three Treatises on the Divine Images* (New York: St. Vladimir's Seminary Press, 2003), 36.

65. See Stanley Hauerwas and Jean Vanier, *Living Gently in a Violent World: The Prophetic Witness of Weakness* (Downers Grove, Ill.: Intervarsity Press, 2008).

The Art of Creaturely Life: A Question of Human Propriety
Norman Wirzba

1. Speaking of sin, Bonhoeffer observes, "the word *disobedience* fails to describe the situation adequately. It is rebellion, the creature's stepping outside of the creature's only possible attitude, the creature's becoming creator, the destruction of creatureliness, a defection, a falling away [*Sturzen*] from being safely held as a creature"; Dietrich Bonhoeffer, *Creation and Fall: A Theological Exposition of Genesis 1–3*, in *Dietrich Bonhoeffer Works*, ed. John W. De Gruchy, trans. Douglas Stephen Bax (Minneapolis: Fortress Press, 1997), 3:120. Subsequent references will be in the text as *CF*.

2. David Kelsey's magisterial two-volume *Eccentric Existence: A Theological Anthropology* (Louisville: Westminster John Knox Press, 2009) is a notable exception to this tendency. Kelsey observes, along with Gustaf Wingren, that the doctrine of creation does "remarkably little work" in modern systematic theology (160). Admittedly, a considerable amount has been written about humanity made in the image of God (*imago Dei*).

What is striking about many of these accounts is how much they rely on philosophical characterizations of capacities such as reason or language or the soul that do not have their inspiration in biblical depictions of creation and creatureliness. For a lucid treatment showing why these accounts are theologically unsatisfactory, see Ian McFarland's *The Divine Image: Envisioning the Invisible God* (Minneapolis: Fortress Press, 2005).

3. Ellen Davis has developed the agrarian context for understanding Israelite history and its theological traditions in *Scripture, Culture, and Agriculture: An Agrarian Reading of the Bible* (New York: Cambridge University Press, 2009).

4. Wendell Berry, *The Unsettling of America: Culture and Agriculture* (San Francisco: Sierra Club Books, 1977), 86.

5. Scripture leaves it open as to whether or not the divine breath that animates the *adam* also animates plant and animal life. The King James Version of the Bible made a clear distinction between human and non-human creaturely life by saying the former became a "living soul," as compared with animals who are "living creatures." The Hebrew, however, does not allow this neat (and entirely advantageous to us) bifurcation, since it names humans *and* animals as *nefesh chaYäh*.

6. The NRSV translation of Genesis 2:15 as "till and keep" clearly resonates with the horticultural context of this passage. Davis argues that the root verb "to work or till" can have a variety of meanings, ranging from working the land to working *for* the land (as a form of service to it, and perhaps even worship to God). The verb "to keep" also has the meaning "to observe" (as when the Israelites are told to observe God's commandments), suggesting that "keeping" presupposes personal alignment or attunement to what is going on and expected in the garden. Davis suggests the translation "And YHWH God took the human and set him in the garden of Eden to work and serve it, to preserve and observe it"; Davis, *Scripture, Culture, and Agriculture*, 30. This translation highlights how *Adam* needs to develop the very practical skills of attention, patient work, and respect for limits and possibilities. The human creature must show humility as one who draws its life from humus. I have developed the meaning of humility in "The Touch of Humility: An Invitation to Creatureliness," in *Modern Theology* 24, no. 2 (April 2008): 225–44.

7. It is important to underscore that gardens are built environments to the extent that they are the coming together of wild and natural forces with human design and skill. But unlike other built environments (a

shopping mall, for instance), in which *human* ingenuity, technology, and ambition dominate and are continually reflected back to us, gardens are places in which people are more readily compelled to see the variety and complexity of creatures and life-processes *beyond* human design or control. The philosopher David Copper describes the deep meaning and attraction of gardens as residing, in part, in their ability to reveal "the relation between the source of the world and ourselves"; Copper, *A Philosophy of Gardens* (Oxford: Clarendon Press, 2006), 150. Speaking of the great variety of gardening traditions around the world, he stresses that attention to the "source of the world" brings us face to face with the mystery of things as present and somehow given: "The Garden, to put it portentously, is an epiphany of man's relationship to mystery. This relationship is its mystery" (145).

8. The poet Rainer Maria Rilke writes in Sonnet 12 of *The Sonnets to Orpheus: First Series*, trans. A. Poulin Jr., *Duino Elegies and the Sonnets to Orpheus* (New York: Mariner, 2005), 107:

> *Selbst wenn sich der Bauer sorgt und handelt,*
> *Wo die Saat in Sommer sich verwandelt,*
> *Reicht er niemals hin. Die Erde schenkt.*

> Though he works and worries, the farmer
> never reaches down to where the seed turns
> into summer. The earth grants.

Wendell Berry speaks similarly in a Sabbath poem from *A Timbered Choir: The Sabbath Poems 1979–1997* (Washington, D.C.: Counterpoint, 1998), 18:

> Whatever is foreseen in joy
> Must be lived out from day to day.
> Vision held open in the dark
> By our ten thousand days of work.
> Harvest will fill the barn; for that
> The hand must ache, the face must sweat.

> And yet no leaf or grain is filled
> By work of ours; the field is tilled

And left to grace. That we may reap,
Great work is done while we're asleep.

9. I have developed this theme in Wirzba, "The Dark Night of the Soil:
An Agrarian Approach to Mystical Life," in *Christianity and Literature*
56, no. 2 (Winter 2007): 253–74.

10. Berry, "A Native Hill," in *The Long-Legged House* (1969; repr.
Washington, D.C.: Shoemaker and Hoard, 2004), 207.

11.

O bent by fear and sorrow, now bend down,

Leave word and argument, be dark and still,
And come into the joy of healing shade.
Rest from your work. Be still and dark until

You grow as unopposing, unafraid
As the young trees, without thought or belief;
Until the shade Sabbath light has made

Shudders, breaks open, shines in every leaf.

A Timbered Choir, 31

12. Bonhoeffer proposes that prior to the transgression Adam lives in
respectful obedience to the grace of life. His obedience made possible
a unified, singularly focused form of life in which the two-sidedness of
good and evil, i.e., the prospect of an option that diverges from obedience,
had not yet emerged. The possibility of good versus evil, therefore, only
emerges *in* the transgressive act. This is why Bonhoeffer thinks Adam first
lives "beyond good and evil" (*CF*, 87). The option is not between good and
evil but between a "life obedient to God" and "a life of good and evil."

13. Carolyn Merchant has described one dimension of modern subjectivity as the scientist who interrogates, even tortures, nature so as to
extract its secrets and bounty. She quotes Francis Bacon: "For like as a
man's disposition is never well known or proved till he be crossed, nor
Proteus ever changed shapes till he was *straitened* and *held fast*, so nature
exhibits herself more clearly under the *trials* and *vexations* of art [mechan-

ical devices] than when left to herself"; Merchant, *The Death of Nature: Women, Ecology and the Scientific Revolution* (San Francisco: Harper, 1980), 169. Rather than being the servants of creation described in Genesis 2, humans are now narrated as the masters of a feminine, constrained, slave-like nature. Pierre Manent, in his examination of modern political thought, casts the gospel of Hobbes, Locke, and Rousseau as such: "In the beginning, the world was without form and void, without laws, arts, or sciences, and the spirit of man moved over the darkness"; Manent, *The City of Man* (Princeton: Princeton University Press, 1998), 183. Modern humanity has become historical, which means it flees God, nature, tradition, and established law. It is a form of constant rebellion in which the only acceptable law is the (arbitrary) law the human being gives to itself.

14. Jean-Luc Nancy, "Introduction," in *Who Comes After the Subject?* ed. Eduardo Cadava, Peter Connor, and Jean-Luc Nancy (New York: Routledge, 1991), 4. Subsequent references will be in the text as *WCAS*.

15. Jean-Louis Chrétien's description of the "call and response" structure of human existence is an essential supplement to Marion's account of the *interloqué*. Before we speak we are always already called by and joined to another: "We speak for having heard. Every voice, hearing without cease, bears many voices within itself because there is no first voice"; Chrétien, *The Call and the Response*, trans. Anne A. Davenport (New York: Fordham University Press, 2004), 1. Insofar as our response is constituted by love, "Our task is not to give an answer that would in some sense erase the initial provocation by corresponding to it, but to offer ourselves up as such in response, without assigning in advance any limit to the gift" (13).

16. Berry observes, "There is, in practice, no such thing as autonomy. Practically, there is only a distinction between responsible and irresponsible dependence"; Berry, *Unsettling of America*, 111.

17. Levinas asks, "Isn't the liveliness of life excessiveness, a rupture of the containing by the uncontainable, a form that ceases to be its proper content already offering itself in the guise of experience—an awakening to consciousness in which the consciousness of awakening is not the truth, an awakening that remains a first movement—a first movement toward the other of which the intersubjective reduction reveals the traumatism, secretly striking the very subjectivity of the subject? Transcendence" (*WCAS*, 215). Answering his own question with transcendence, Levinas makes clear that the fundamental question is how to live in the face of

limit without transgressing and appropriating it. Genuine life is excessiveness and amazement before what comes to me from beyond myself.

18. Bonhoeffer succumbs, incorrectly in my view, to a history of interpretation in which the naming of animals is equated with mastery over
them. He says of the animals, "They remained a strange world to Adam;
indeed they remain, for all their nature as siblings, creatures subjected to,
named by, and ruled over by, Adam" (*CF*, 96–97). Clearly there are forms
of naming that do establish hierarchies and systems of domination, but
the biblical text does not in the first instance warrant this interpretation.
Adam's naming takes place before sin has entered into and distorted relationships. It makes more sense to say that naming makes possible relationships, the precise character of which is yet to be determined. Naming
one way or another simply lays out different ways of relating to others (for
example, naming a plant a "fruit," a "flower," or a "weed" evokes different responses within us).

19. Bonhoeffer describes shame as expressing the fact "that we no longer
accept the other as God's gift but instead are consumed with an obsessive
desire for the other. . . . Shame is a cover in which I hide myself from the
other because my own evil and the other person's evil, that is, because of the
dividedness that has come between us" (*CF*, 101). One can, in this regard,
compare Levinas, who describes shame as the freedom that has become
murderous; Levinas, *Totality and Infinity: An Essay on Exteriority*, trans.
Alphonso Lingis (Pittsburgh: Duquesne University Press, 1969), 83–84.

20. Erazim Kohák, *The Embers and the Stars: A Philosophical Inquiry
into the Moral Sense of Nature* (Chicago: University of Chicago Press,
1984), 107. Kohák continues: "the living truth of having is belonging, the
bond of love and respect which grows between one being and another in the
course of seasons. The claim to having is as strong as all the love and care a
person gives, and only that strong. It is crucial to have no more than we can
love, for without love the claim to having becomes void. Loveless having,
possessing in the purest sense, remains illegitimate, a theft" (107–8).

21. Bonhoeffer develops this theme in *Ethics*, vol. 6 of *Dietrich Bonhoeffer Works*, ed. Clifford Green, trans. Reinhard Krauss, Charles C.
West, and Douglas W. Stott (Minneapolis: Fortress Press, 2005), 306–7.

22. In *Letters and Papers from Prison*, vol. 8 of *Dietrich Bonhoeffer
Works*, ed. John De Gruchy, trans. Isabel Best, Lisa E. Dahill, Reinhard
Krauss, and Nancy Lukens (Minneapolis: Fortress Press, 2009), Bonhoeffer

wrote, "one only learns to have faith by living in the full this-worldliness of life. If one has completely renounced making something of oneself . . . then one throws oneself completely into the arms of God, and this is what I call this-worldliness: living fully in the midst of life's tasks, questions, successes and failures, experiences and perplexities—then one takes seriously no longer one's own sufferings but rather the sufferings of God in the world. Then one stays awake with Christ in Gethsemane. And I think this is faith; this is *metanoia*. And this is how one becomes a human being, a Christian" (486).

23. Berry describes how in many cultures it has been important for the journey to adulthood to also be a journey through wilderness so that people might come to an appreciation of their place within creation. "Seeing himself as a tiny member of a world he cannot comprehend or master or in any final sense possess, he cannot possibly think of himself as a god. And by the same token, since he shares in, depends upon, and is graced by all of which he is a part, neither can he become a fiend; he cannot descend into the final despair of destructiveness. Returning from the wilderness, he becomes a restorer of order, a preserver. . . . He embodies the passing of human time, living and dying within the human limits of grief and joy"; Berry, *Unsettling of America*, 99.

24. Numerous historical accounts that chart the modern development of the metaphor of the world as a machine are available. One can begin with R. G. Collingwood's classic treatment in *The Idea of Nature* (Oxford: Oxford University Press, 1960), and then move to Merchant's *Death of Nature* for a brief overview.

25. "Both in principle and as a matter of historical fact, alienation sets in when humans lose their awareness of the presence of God and persuade themselves to view the cosmos no longer as a creation, endowed with value in the order of being, a purpose in the order of time and a moral sense in the order of eternity, but as a cosmic accident, meaningless and mechanical. Then nature comes to appear as absurd and we ourselves as futile within it. In fact, we first desanctified nature by exiling God into the 'supernatural'"; Kohák, *Embers and the Stars*, 183. Kohák argues that there is no acknowledgment more primordial than the embodied experience of being held and nurtured—"at home"—in a life-supporting world. Daily life demonstrates that we belong in creation. Our belonging, however, does not render our created home as comprehensible or as

always convenient and comfortable. A problem with so much modern science and technology is that they deny our experience of belonging and replace it with a theoretical, objective construct that ultimately renders people bored or apathetic. What I mean by this can be seen in the rise of persons who describe themselves as "apatheists" rather than theists or atheists. *USA Today*, in a news story on growing spiritual apathy among Americans, quotes Ben Helton, a high school band teacher in Chicago, who says, after considering what modern science teaches about evolutionary psychology, "we might as well be cars. That, to me, makes more sense than believing what you can't see" (Tuesday, January 3, 2012, page 9A). This view of persons as cars, a view perfectly suited to (and perhaps made inevitable by) a consumeristic world, presupposes that we are fundamentally without value, isolated, and finally incapable of love.

26. Henry Bugbee, *The Inward Morning: A Philosophical Exploration in Journal Form* (1958; repr. Athens: University of Georgia Press, 1999), 76. Bugbee maintains that our experience has been rendered shallow by being packaged and stylized by scientists, philosophers, and marketers: "experience is a tissue of meaning grossly misinterpreted by representation in the image of the object. . . . Experience is our undergoing, our involvement in the world, our lending or withholding of ourselves, keyed to our responsiveness, our sensibility, our alertness or our deadness" (41). In a spectator, consumerist culture, "experiences" are "events" to be consumed.

27. Thomas A. Carlson. *The Indiscrete Image: Infinitude and Creation of the Human* (Chicago: University of Chicago Press, 2008), 29. Subsequent references will be in the text as *TII*.

28. Jacques Derrida, *Given Time: 1. Counterfeit Money*, trans. Peggy Kamuf (Chicago: University of Chicago Press, 1992), 12. Derrida continues, "For there to be gift, it is necessary that the gift not even appear, that it not be perceived or received as a gift" (16).

29. Robyn Horner, *Rethinking God as Gift: Marion, Derrida, and the Limits of Phenomenology* (New York: Fordham University Press, 2001), 18.

30. I have developed the sense of mystery and awe that accompanies eating in *Food and Faith: A Theology of Eating* (New York: Cambridge University Press, 2011).

31. See Mark Manolopoulos's *If Creation Is Gift* (Albany: SUNY Press, 2009) for a treatment of the gift aporia in light of ecological degradation.

32. This giving of oneself is not a giving that follows from obligation or debt, because the very idea of debt presupposes a ledger in which credits and debits can be clearly delineated. The depth, breadth, and mystery of creaturely membership exceed what any such ledger could possibly contain. Horner is correct, therefore, to say, "if I give, it can only be because I feel I have been gifted with the capacity to give, not because I feel that I must give back"; Horner, *Rethinking God as Gift*, 183.

33. In "Aspects of a Doctrine of Creation," in *The Doctrine of Creation: Essays in Dogmatics, History, and Philosophy*, ed. Colin Gunton (Edinburgh: T. and T. Clark, 1997), Robert Jenson, drawing on the insight of John of Damascus, develops the theme of creation as the work of a hospitable God making room for others within the triune life.

34. Rowan Williams, "Creation, Creativity, and Creatureliness: The Wisdom of Finite Existence," a lecture delivered April 23, 2005, at the St. Theosevia Centre for Christian Spirituality in Oxford (http://www.archbishopofcanterbury.org/articles.php/2106/creation-creativity-and-creatureliness-the-wisdom-of-finite-existence). This essay is included in this volume as Chapter 1.

35. Berry, *Life Is a Miracle: An Essay Against Modern Superstition* (Washington, D.C.: Counterpoint, 2000), 137–38. Berry elaborates: "I don't think creatures can be explained. . . . What we know about creatures and lives must be pictured or told or sung or danced. . . . The arts are indispensable precisely because they are so nearly antithetical to explanation" (113).

36. Kohák argues that it is important to alleviate the pain of others insofar as we can. But it is also crucial for us to learn to live with pain as a learning to live within limits. "Pain borne and shared . . . teaches the human his own insufficiency, his own need and, with it, gentleness. It opens him to receive, in empathy, the gift of the other, not in censure but in gratitude and love"; Kohák, *Embers and the Stars*, 46.

Face of Nature, Gift of Creation: Thoughts Toward a Phenomenology of *Ktisis*
Bruce Foltz

1. Martin Heidegger, *Elucidations of Hölderlin's Poetry*, trans. Keith Hoeller (Amherst, N.Y.: Humanity Press, 2000), 78ff.

2. Bill Devall and George Sessions, *Deep Ecology: Living as If Nature Mattered* (Salt Lake City: Peregrine Smith, 1985), 98–100.

3. Heidegger, *The Fundamental Concepts of Metaphysics: World, Finitude, Solitude*, trans. William McNeill and Nicholas Walker (Bloomington: Indiana University Press, 1995), 263.

4. Heidegger, *Elucidations*, 69.

5. Ibid., 79.

6. Ibid., 69.

7. Hubert Dreyfus and Sean Dorrance Kelly, *All Things Shining: Reading the Western Classics to Find Meaning in a Secular Age* (New York: Free Press, 2011), xi.

8. Ibid.

9. Ibid., 222ff.

10. Frederich Schiller, *On the Sublime: Two Essays*, trans. Julius A. Elias (New York: Fredrick Ungar, 1967), 120.

11. Albert Camus, *Lyrical and Critical Essays*, ed. Ellen Conroy Kennedy (New York: Random House, 1970), 150ff.

12. Gerard Naddaf, *The Greek Concept of Nature* (Albany: SUNY Press, 2005) 34f., 168ff.

13. Erazim Kohák, *The Embers and the Stars: A Philosophical Inquiry into the Moral Sense of Nature* (Chicago: University of Chicago Press, 1984), 68ff., italics added.

14. This draws upon two different translations: Carl A. P. Ruck and William H. Matheson, *Selected Pindar Odes* (Ann Arbor: University of Michigan Press, 1968), 63, and C. M. Bowra, *The Odes of Pindar* (Baltimore: Penguin, 1968), 165.

15. *The Psalter According to the Seventy* (Boston: Holy Transfiguration Monastery, 1997), 186ff.

16. This passage from St. Cyril's *Catechetical Lectures* is cited in Quentin F. Wesselschmidt, *Ancient Christian Commentary on Scripture: Old Testament VII, Psalms 51—150* (Downer's Grove, Ill.: Intervarsity Press, 2007), 236.

17. Ibid.

18. Heidegger, *An Introduction to Metaphysics*, trans. Ralph Mannheim (New Haven: Yale University Press, 1957), 14; translation altered.

19. Ibid.; italics added.

20. Heidegger, *Nietzsche: Volumes Three and Four*, trans. David Krell (New York: Harper Collins, 1991), 7.

21. Both St. Isaac and St. Maximus are cited here from Olivier Clément, *The Roots of Christian Mysticism* (London: New City Press, 1993), 213, 217.

22. Isa. 6:3; Athanasius, *On the Incarnation of the Word*, Section XLV, in *Nicene and Post-Nicene Fathers*, vol. 4, *Athanasius: Select Works and Letters*, edited by Philip Schaff, 61 (Grand Rapids: Eerdmans, 1987), 61.

23. See Sergei Bulgakov, *The Lamb of God*, trans. Boris Jakim (Grand Rapids: Eerdmans, 2008) 164.

24. Pavel Florensky, *The Pillar and Ground of the Truth: An Essay in Orthodox Theodicy*, trans. Boris Jakim (Princeton: Princeton University Press, 1997) 521, 202.

25. Translated in Florensky, *Pillar*, 200.

26. St. Nikodimos of the Holy Mountain and St. Makarios of Corinth, *The Philokalia: The Complete Text*, trans. G. E. H. Palmer, Philip Sherrard, and Kallistos Ware (London: Faber and Faber, 1995), 4:92.

27. Citations from St. Anthony and St. Gregory, in Florensky, *Pillar*, 526ff.

28. Florensky, *Pillar*, 192.

29. Fyodor Dostoevsky, *The Brothers Karamozov*, trans. Richard Pevear and Larissa Volokhonsky (New York: Random House, 2000), 288ff.

30. Ibid., 298ff.

31. Ibid., 294ff.; italics added.

32. St. John of Damascus, *Three Treatises on the Divine Icons*, trans. Andrew Louth (Crestwood, N.Y.: St. Vladimir's Seminary Press, 2003), Kindle ed., locations 348–52.

33. Cited in Pierre Hadot, *What Is Ancient Philosophy?*, trans. Michael Chase (Cambridge, Mass.: Harvard University Press, 2002), 238.

34. Thomas Traherne, *The Centuries*, in *Centuries, Poems, and Thanksgivings*, vol. 1, *Introduction and Centuries*, ed. H. M. Margoliouth (London: Oxford University Press, 1958), 3.

35. Ibid., 3ff. (1:3)

36. Ibid., 33. (1:64)

37. Ibid., 50. (1:92)

38. Ibid., 3. (1:2)

39. Ibid., 15. (1:31)

40. Ibid., 3. (1:1)

41. Ibid., 15. (1:29)

42. John Muir, *The Cruise of the Corwin: Journal of the Arctic Expedition of 1881 in Search of DeLong and the Jeannette*, ed. William Frederic

Badè, in *John Muir: His Life and Letters and Other Writings*, ed. Terry Gifford (London: Bâton Wicks, 1996), 746; italics added.

43. Florensky, *Iconostasis*, trans. Donald Sheehan and Olga Andrejev (Crestwood, N.Y.: St. Vladimir's Seminary Press, 1996), 50ff.; italics added.

44. Ibid.

45. Ibid., 52.

46. Christos Yannaras, *Person and Eros*, trans. Norman Russell (Brookline, Mass.: Holy Cross Orthodox Press, 2007), 50.

47. Bulgakov, *The Comforter*, trans. Boris Jakim (Grand Rapids: Eerdmans, 2004), 188.

48. Bulgakov, *The Bride of the Lamb*, trans. Boris Jakim (Grand Rapids: Eerdmans, 2002), 402.

49. Dostoevsky, *Demons*, trans. Richard Pevear and Larissa Volokhonsky (New York: Random House, 1994), 145.

50. Bulgakov, *The Burning Bush*, trans. Boris Jakim (Grand Rapids: Eerdmans, 2009); *Burning Bush*, Kindle ed., location 2101.

Creativity as Call to Care for Creation?
John Zizioulas and Jean-Louis Chrétien
Christina M. Gschwandtner

1. Jean-Louis Chrétien, *The Ark of Speech*, trans. Andrew Brown (London: Routledge, 2004), 124. The quotation from Philo is found on the same page.

2. This claim is made fairly frequently by Patriarch Bartholomew, for example, but also by various other Orthodox thinkers. For a summary of these contemporary claims, see Elizabeth Theokritoff, *Living in God's Creation: Orthodox Perspectives on Ecology* (Crestwood, N.Y.: St. Vladimir's Seminary Press, 2009), 211–52, where she summarizes the contemporary emphasis on human ecological priesthood together with some other contemporary Orthodox approaches.

3. For the strongest statement of this, see John Zizioulas, "Proprietor or Priest of Creation?" *Keynote Address of the Fifth Symposium of Religion, Science, and the Environment* (2 June 2003), http://www.orthodoxy today.org/articles2/MetJohnCreation.php.

4. For the most prominent exposition of this, see the works of Douglas J. Hall: *Imaging God: Dominion as Stewardship* (Grand Rapids: Eerd-

mans, 1986); *The Steward: A Biblical Symbol Come of Age* (Grand Rapids: Eerdmans, 1990); and *Professing the Faith: Christian Theology in a North American Context* (Minneapolis: Fortress Press, 1993).

5. It goes without saying that most major theological doctrines also circle around the human: God becomes human in Christ, redeems humans, and promises the eschaton to humans.

6. Human uniqueness has often been employed to justify human superiority over the rest of creation in a way that permits exploitation. The idea that humans are absolutely distinct from and thus superior to the rest of creation has been challenged in recent years as one of the "enabling factors" for the ecological crisis. It is claimed that regarding humans as somehow special, as other than animals or plants, leads directly to license to denigrate and ruthlessly exploit those animals and plants. This was first articulated by Lynn White in his seminal (though rather general) essay "The Historical Roots of the Ecologic Crisis"(*Science* 155 [1967]: 1203–7), where he made (Western) Christianity responsible for the crisis because of God's command in Genesis to "subdue and dominate" creation. The argument has been taken up since then in many shapes and forms. Many environmental philosophers say that we must instead propose a more egalitarian environmental ethic that recognizes that all living beings are an intricate part of the cycle of life and that humans hold no special place in it. This idea is especially put forward by deep ecology and some other radical movements. Most approaches in environmental ethics on some level combat anthropocentrism as a key problem, but there is much controversy over how "egalitarian" our ethic should be and what this means in terms of assigning moral value to nonhuman beings.

7. In fact, it appears there far more often and is far more fully discussed than in Western sources. It is a topic that occupies almost every Patristic thinker, and contemporary introductions to Orthodox thought usually also devote a section to it.

8. This idea has a very rich tradition that cannot be discussed in any detail here. There are individual studies of the notion of the *imago dei* in most Patristic thinkers. It is a central notion in the Orthodox theological tradition, both past and present, although there is no definite agreement on how this idea is to be interpreted; for a brief summary of the idea and its connection to "dominion," see Theokritoff, *Living in God's Creation*, 70–73. Many Eastern thinkers also draw a distinction between "image"

and "likeness," some suggesting that the image is part of the very definition of the human and retained after the fall, while the likeness is to be
regained through moral (especially ascetic) effort that leads to theosis/
deification—i.e., likeness with the divine, a central Orthodox soteriologi-
cal doctrine. For an excellent discussion of the *imago dei* in Patristic
literature, see T. Camelot, "La théologie de l'image de Dieu," *Revue des
sciences philosophique et théologiques* 40 (1956): 443–65.

9. Ernst Conradie examines this notion in great detail and criticizes
its ecological repercussions. He seeks to establish a new more ecologically
oriented anthropology in Conradie, *An Ecological Christian Anthropology:
At Home on Earth?* (Aldershot: Ashgate, 2005).

10. Patristic thinkers frequently stress that everything in the world
was created for human use (and thus often go to great trouble in explaining how certain aspects of creation could possibly be "useful" to humans).
Even twentieth-century Rumanian Orthodox theologian Dumitru Staniloae claims that "the world has no meaning, if it is not understood as the
gift of God to mankind"; Staniloae, "The World as Gift and Sacrament of
God's Love," *Sobornost* 5, no. 9 (1969): 665.

11. This idea is also first explicitly articulated by Plato. It has become
most popularly associated with Maximus the Confessor, although it is
used by many Eastern Patristic thinkers (Gregory of Nyssa is one of the
few who is somewhat critical of it). For a summary of the use of Maximus
in Orthodox environmental reflection, see Theokritoff, *Living in God's
Creation*, 53–67. She also points out that extrapolating Maximus's idea
of the microcosm into a notion of "priesthood" on behalf of creation is a
more recent phenomenon in Orthodox thinking (216).

12. Plants, animals, and inanimate matter are only physical or material
(they have no souls), while angels and the divine are spiritual and do not
have matter. Humans have both, thus participating in both spheres, and
can serve as the connecting link between them.

13. This is explicated most fully in Maximus's *Mystagogy* and in Dif-
ficulty 10 of his earlier collection of *Ambigua* (*Book of Difficulties*). For a
translation and a useful introduction to Maximus's thought, see Andrew
Louth, *Maximus the Confessor* (London and New York: Routledge, 1996).

14. In his book *Communion and Otherness: Further Studies on Personhood and the Church* (London: T. and T. Clark, 2006), which collects
several of his earlier essays, Zizioulas articulates his stance on the human

role in the larger creation more fully and more carefully than in his many briefer addresses on ecological issues. For his more explicitly ecological addresses and articles (which almost all focus on the idea that humans are to serve as priests of creation), see Zizioulas, "Orthodoxy and Ecological Problems: A Theological Approach," in *The Environment and Religious Education: Presentations and Reports, Summer Session on Halki 1994*, ed. D. Tarasios Istanbul: Melitos, 1997); "Ethics Versus Ethos: An Orthodox Approach to the Relation between Ecology and Ethics," in *The Environment and Ethics: Presentations and Reports, Summer Session on Halki 1995*, ed. D. Tarasios (Istanbul: Melitos, 1997); "Preserving God's Creation: Three Lectures on Theology and Ecology," *King's Theological Review* 12, no. 1 (1989): 1–5; 12, no. 2 (1989): 41–45; and 13, no. 1 (1990): 1–5; and "Proprietor or Priest of Creation?" *Keynote Address of the Fifth Symposium of Religion, Science, and the Environment* (2 June 2003), http://www.orthodoxytoday.org/articles2/MetJohnCreation.php. For a summary of Ecumenical Patriarch Bartholomew's writings and speeches on the environment, see Christina M. Gschwandtner, "Orthodox Ecological Theology: Bartholomew I and Orthodox Contributions to the Ecological Debate," *International Journal for the Study of the Christian Church* 10, nos. 2–3 (2010): 130–43.

15. See the discussion in Zizioulas, *Communion and Otherness*, 39.

16. Ibid., 40; emphasis mine.

17. This idea is also put forward by other contemporary Orthodox thinkers, such as Kallistos Ware, Christos Yannaras, and Patriarch Bartholomew. For one example, see Ware's chapter in *Living Orthodoxy in the Modern World*, ed. Andrew Walker and Costa Carras (London: SPCK: 1996).

18. Zizioulas, *Communion and Otherness*, 41.

19. "In our time, when human *nature* is, as we have seen, hardly distinguishable from the rest of mammals by biological science, it is the language of *call* rather than substance that would express better the emergence of the human being from the 'hands' of the creator"; Zizioulas, *Communion and Otherness*, 42.

20. The importance of relationality is central to Zizioulas's thought and was set forth already in his early text *Being and Communion: Studies in Personhood and the Church* (Crestwood, N.Y.: St. Vladimir's Seminary Press, 1985). In *Communion and Otherness* he highlights how this helps

distinguish humans from other animals: "By making the human being emerge as a particularity in creation through the divine call, we are defining it as a being distinguished from the rest of the animals not genetically, but by way of *relationship* to God and the rest of creation, that is, by its *freedom*" (42).

21. There are interesting parallels on this point between the work of John Zizioulas and contemporary French philosopher Emmanuel Falque. Falque also interprets the fall as a return to bestiality (and a reduction of our animality to mere bestiality); see especially his *Les noces de l'agneau: Essai philosophiques sur le corps et l'eucharistie* (Paris: Cerf, 2011).

22. Zizioulas draws on twentieth-century philosophy, from Husserl and Heidegger, via Sartre, to Buber and Lévinas in order to articulate a self as other, while criticizing them in light of insights from Patristic thinkers. Some of this critique is a bit superficial, especially in regard to Lévinas, whom he uses the most explicitly and criticizes the most vigorously. He only uses early Lévinassian texts, and much of his critique would be alleviated if *Otherwise Than Being* and other later texts were examined more fully.

23. Zizioulas, *Communion and Otherness*, 40.

24. Philip Sherrard, one of the first Orthodox writers to address environmental issues, argues that humans have a special calling to make the rest of creation sacred and concludes, "It is on us that creation depends. It is in our life that creation lives"; Sherrard, *The Sacred in Life and Art* (Ispwich, UK: Golgonoza Press, 1990), 13. Kallistos Ware, in an address devoted to environmental issues, also emphasizes this unique role: "Through our power of self-awareness, and through this ability to alter and restructure the world, we humans are able to give creation a tongue, rendering it eloquent in praise of God"; Ware, *Through the Creation to the Creator* (London: Friends of the Centre Papers, 1997), 23. Various other Orthodox thinkers now articulate something of this sort in their writings.

25. The terminology of "hypostasis" goes back to the early Trinitarian controversies and was originally employed to distinguish the three divine persons within the Trinitarian *ousia*. Both *ousia* and hypostasis are ways of designating being: hypostasis was used to express particular being or nature—i.e., fatherhood, sonhood, and spiritness—and *ousia* to express the general being or nature—i.e., divinity. Yet it is only much more recently that hypostasis has been applied to human beings and even more recently

that it is taken to express an activity: humans are to "hypostasize" the rest of creation in their person. One of the first to articulate this (and certainly the person to explicate it the most fully) was the early-twentieth-century Russian Orthodox theologian Sergei Bulgakov (probably the most original and most important Orthodox theologian since Gregory Palamas, although many Orthodox object to his doctrine of sophiology and some other aspects of his thinking). Bulgakov explicitly speaks of the human task of hypostasizing or humanizing the rest of creation and interprets this as a call to improve the world through creativity, but also through industrialization and labor. He claims that humans hypostasize animals through domestication and farming practices. Overall, he has a fairly positive view of technology and industrialization, as is also true of other twentieth-century Orthodox thinkers (such as the important Rumanian theologian Dimitru Staniloae; the Serbian theologian Justin Popovitch is an interesting exception to this trend). This interpretation of human relation to the rest of creation through their domestication or agricultural use is of course deeply problematic and exacerbates interpretations of the *imago dei* in terms of rule over other creatures.

26. "Love as eros hypostasizes beings, that is, makes them exist as particular, by incorporating them into a unique (a uniquely loved) hypostasis. Not only does uniqueness not exclude other particular beings, it establishes their otherness in and through communion"; Zizioulas, *Communion and Otherness*, 74.

27. This is obviously true of most other theologians and philosophers, not only of Zizioulas.

28. "The Church is the community in which otherness is experienced as communion in and through uniqueness"; Zizioulas, *Communion and Otherness*, 75.

29. Ibid., 92.

30. Ibid., 93. In this context he criticizes Heidegger's thinking as a sort of pantheistic paganism and claims that the Eucharistic approach he proposes is able to strike a balance between a pure objectification of nature and a false subjectification that sees nature as sacred in itself (93–95). He explicitly states here that only humans have been given the gift of freedom, and thus they alone are persons (95).

31. Ibid., 95, 96. This is rather ironic in light of the fact that the unprecedented present rate of elimination of living species is due almost entirely to human activity.

32. He claims that Byzantine art portrays nature in its "eschatological" sense (which seems to equate with its absence). "Just as in the Eucharist, in the hands of the human beings, the material creation is liberated from mortality by becoming the 'body of Christ' and by being drawn into the personal relationship of Christ with his Father in the Holy Spirit, in a similar manner art liberates nature and matter from death by granting them 'eternal life' through the personal seal of the creator. . . . The eucharistic ethos leads to art as a truly *creative* manifestation of personhood, as the emergence of otherness in an ontological and personal sense"; ibid. A curious combat between freedom and necessity runs through all of Zizioulas's work. Love and creativity are signs of personhood and the result of freedom, while nature is entirely tied to fate and necessity.

33. Zizioulas, "Man the Priest of Creation: A Response to the Ecological Problem," in Walker and Carras, *Living Orthodoxy in the Modern World*, 183; emphases mine.

34. Indeed, Ruth Page suggests that this idea limits God, who apparently is unable to have a direct relationship with creation and instead requires human mediation; Page, *God and the Web of Creation* (London: SCM Press, 1996), 162. Conradie is also quite critical of this idea; *Ecological Christian Anthropology*, 210.

35. For example, did God have to "wait" for billions of years for humans to emerge in order to enjoy the dinosaurs and other creatures now extinct? Or, as God is not in time in the way we are, does our liturgical praise today somehow help the divine enjoy the dinosaurs and other animals that became extinct millions of years before we emerged on the scene?

36. See for example, Bulgakov, *The Bride of the Lamb* (Grand Rapids: Eerdmans, 2008), 149–50; Bulgakov, *The Comforter* (Grand Rapids: Eerdmans, 2004), 189–358; and Bulgakov, "Orthodoxy and Economy," in *The Orthodox Church*, trans. E. S. Cram (London: Centenary Press, 1935), 193–97.

37. See, for example, Staniloae's article, "The World as Gift and Sacrament of God's Love," *Sobornost* 5, no. 9 (1969): 662–73, where he suggests that economic labor and civilization are a way of pleasing God and participating in the "progress of creation."

38. This seems to become particularly problematic when the human role is identified with that of the divine. Humans are creators inasmuch as they imitate the divine creation. They receive a unique call from God

that no other creature can hear. They alone can "lift" up the creation (in imitation of Christ) and give a voice to it. Only humans are truly liturgical actors; the rest of creation becomes a mere object for liturgy. These claims are exacerbated by a dangerous tendency in much contemporary Orthodox theology to equate human and divine, to regard humans as quasi-divine. The fundamental Patristic dividing line between the uncreated (i.e., God) and the created (i.e., everything else, including angels and humans) seems to have shifted to include the human on the side of the divine standing over and against what is nonhuman. Thus Alfeyev speaks of humans as partaking "of the divine nature by the very act of creation" such that they "are thereby utterly different from other living beings." Humans are "demi-gods," "commensurable, almost 'identical' to God"; Hilarion Alfeyev, *The Mystery of Faith: An Introduction to the Teaching and Spirituality of the Orthodox Church* (London: DLT, 2002), 59, 64. This idea is also especially put forward by Sophrony Sakharov, who founded one of the first Western Orthodox monasteries after the Russian emigration (in Essex, UK) and whose writings have (unfortunately) become incredibly popular. He speaks of an identity between humans and God and attributes traditional divine attributes such as eternity, omnipresence, omniscience, and omnipotence to humans. This seems deeply problematic to me and not only for its ecological implications.

39. Interestingly enough, in Chrétien this is not within the context of any sort of conscious environmental argument.

40. Chrétien, *Ark of Speech*, 1.

41. The title of this text and the book as a whole play on the reference to Noah's ark, but also to the "arc" of the rainbow of peace, the reaching of an arc that connects things, and a sheltering arch or archway.

42. Chrétien, *Ark of Speech*, 124, 125.

43. Ibid., 114.

44. Ibid., 79. Chrétien's most recent work, *Reconnaissances philosophiques*, part 2, "On the Philosophy of Art and of Beauty" (Paris: Cerf, 2010), 191–308, also contains several essays on the topic of beauty, including an essay on the "life, death and resurrection" of works of art, one on music, one on Kant, one on Joseph Joubert, and one on Hans Urs von Balthasar.

45. "Beauty says adieu when the excess of its manifestation calls out, in our own voice which it causes to falter and seizes with urgency, a name higher than all names, which we alone, albeit in fear and trembling, can

pronounce. This adieu is an *envoi*: we cannot in God himself contemplate the beauty of things created, for of this beauty we are only the lieutenants and the only mouthpieces. It is our voice that carries to God and into God the adieu of things and of the mute world. If things in all their radiance are charged with a mission to us, we too have the mission of welcoming them, gathering them in, carrying them by means of our voice to a place that they cannot reach of themselves"; Chrétien, *Ark of Speech*, 98.

46. Ibid., 131, 132. Chrétien also explicates this poetic singing of the world in three essays on Thomas Traherne, Walt Whitman, and Paul Claudel, respectively, as the three final chapters of Chrétien, *La joie spacieuse: Essai sur la dilatation* (Paris: Minuit, 2007).

47. Chrétien, *The Call and the Response*, trans. Anne Davenport (New York: Fordham University Press, 2004), 9.

48. Ibid., 10.

49. "The fact that my very being is the advent of a response shaped by the call's own scission means that there has never been a first instant of response, that I never started to speak in order to answer. Every initiative on my part only perpetuates an immemorial yes, in the rift between two forms of excess. Infinite excess, first of all, of the call over the answer, since the call is of the infinite: by calling me as a person, it calls me not as an isolated and abstract being but calls the totality of the world in space and time along with me, in the inexhaustible chorus of which I am only one voice enduring a perpetual inchoation"; Anne Davenport, trans., introduction, in Chrétien, *Call and Response*, 19.

50. These points are made especially in the first chapter of his book on response and responsibility. Their intrinsic connection is explored throughout Chrétien's book *Répondre: Figures de la réponse et de la responsabilité* (Paris: PUF, 2007), 2–3. Chrétien makes clear that the call can proceed from things and situations, not only from God or other humans (3–4). Somewhat later in the chapter Chrétien reports the story of an old man he encountered as a young child whose ardent embrace of a tree after the atrocities of the war affirmed humanity in light of the mutilating horrors of war. Although here the tree seems to stand as a mere symbol for saving "the humanity of the human," Chrétien concludes this account by affirming that this shows that things can call and respond, as well (11). See also the further discussion of how things can call to us on pages 12–13 of that work.

51. Thus both go far beyond their Lévinassian inspiration.

52. This articulated more fully both in Chrétien's *The Unforgettable and the Unhoped For*, trans. Jeffrey Bloechl (New York: Fordham University Press, 2002) and in his more recent book on the response, *Répondre: Figures de la réponse et de la responsabilité*.

53. The phenomenology of the call is central to Chrétien's philosophy overall and is articulated in many of his works.

54. Chrétien, *Ark of Speech*, 136.

55. Ibid., 137. This tendency is also evident in his essay on Thomas Traherne, although it is not entirely clear whether he is merely presenting Traherne's views here or explicitly approving of them; Chrétien, *La joie spacieuse*, 173–204, especially 179–83. Yet even here the stress on the uniqueness of the self still is in the context of responsibility and a real call proceeding from nature.

56. Chrétien, *Ark of Speech*, 140; translation slightly modified.

57. "The Holy Bible itself speaks of God only by speaking of the world, of its creation, its history, the human beings who live in it, the animals, the rivers, the mountains and the deserts, the winds and the bushes, of God giving the world and giving himself for the salvation of the world"; Chrétien, *Ark of Speech*, 120.

58. Ibid., 129.

59. Ibid., 132.

60. Chrétien, *L'effroi du beau* (Paris: Cerf, 2008), 71.

61. In fact, in the penultimate chapter of *Hand to Hand: Listening to the Work of Art*, Chrétien wonders precisely about the reversal of artistry and creation that depicts God as a supreme (quasi-human) artist and attributes creation—originally God's sole prerogative—to humans. In the chapter "From God the Artist to Man the Creator," he traces the history of this development; Chrétien, *Hand to Hand: Listening to the Work of Art*, trans. Stephen E. Lewis (New York: Fordham University Press, 2003), 94–129. He points out that turning God into an artist makes nature secondary to the work of art (99). The idea that creation speaks to us is also part of this legacy (105). Although Chrétien is primarily concerned with tracing this history, he does censure a conflation of human and divine making several times in the essay. For example, he says that "to take God as the point of departure to ponder art is, even if the word is not uttered, to consider man as creator, and to tend to efface the dissemblance between

his art and divine art" (111). He also asserts that "to sing the song of the real world is a higher and more dignified task than the creation of fictional works" (115). Eventually human making becomes equivalent to divine creation and divine creation becomes described as a craftsmanship similar to that of human work. Chrétien concludes, "In taking himself for God, man becomes an idle god, capricious and arbitrary" (124). We must recover both the uniqueness of divine creation through the word and the manual labor of the hand involved in all human making, which should not be called "creation" (124–25).

62. Chrétien, L'effroi du beau, 73.

63. Chrétien, Ark of Speech, 130; L'effroi du beau, 76.

64. Chrétien, Ark of Speech, 5. He also warns, "When speech tries to unveil God instead of letting itself be unveiled and stripped bare, the loss of its dramatic tension is also a loss of light and intelligibility. If there is a divine beauty, we cannot be its spectators, but only its witnesses. And the witnesses of such beauty speak in proportion to their wound and their sense of being torn apart" (85).

65. Ibid., 124.

66. See especially the two chapters on humility in Chrétien, Le regard de l'amour (Paris: Brouwer, 2000), 11–54. Although the first sentence of the earlier chapter points out the link between humility and the humus of the earth at the origin of the word, the chapter focuses primarily on our humility vis-à-vis other humans, including a discussion of how it is grounded in creation in the image of God (27–28). The second chapter discusses the theme of humility and its link to love in St. Bernard of Clairvaux. The topic of humility is pursued also in several of the subsequent chapters. For a more general reflection on the significance of humility in light of Chrétien's work, see Norman Wirzba, "The Touch of Humility: An Invitation to Creatureliness," Modern Theology 24, no. 2 (2008): 225–44.

67. Chrétien, Call and Response, 43.

68. "The call that recalls us is also a promise that keeps us beholden; it gives us speech only by gripping us by the throat. No hymn will be able to keep it. Yet every hymn, torn and heartrending, must remit itself to this promise for safekeeping, entrust itself to it, give itself to it and lose itself, always already, always more, never enough"; Chrétien, Call and Response, 32.

69. Chrétien, Ark of Speech, 110.

70. Chrétien, *L'effroi du beau*, 32. This is connected here already with the recognition that beauty comes to us instead of being solely created by us: "Beauty gives a plentitude which does not imply any fulfillment, because this plentitude is its own, not ours. Nothing that is empty in us can measure up to this plentitude, and if there is a desire for beauty, it is only opened in us by beauty itself and belongs to its offering, far from being prior to it. We can suffer from beauty and before it, but we are not in pain from beauty in the sense that we would have been able to miss or lack it. It only fills those it has first opened. . . . This is why beauty comes as a *shock* [effroi]." Chrétien works out a similar notion of "shock" proceeding from beauty highlighting the precariousness of the human condition in subsequent essays in the same book.

71. This is not to say, of course, that the Orthodox theological tradition does not stress human creatureliness. This is in fact a strong theme in the Patristic tradition, which always draws the fundamental distinction between Creator and creatures (instead of between spiritual and material) but seems to recede in contemporary writers, who often place humans on the divine side of the divide instead of emphasizing their creatureliness. Kenosis, modeled on Christ's self-emptying and self-sacrifice, is an important aspect of the Orthodox notion of theosis (deification), as is the virtue of humility.

72. It is important to stress, however, that rejecting *complete* distinction of the human from the rest of creation does not imply that *no* distinction is possible or that there are no differences between creatures. As Conradie points out, we must learn to articulate human difference *among* creatures instead of *from* creatures; Conradie, *Ecological Christian Anthropology*, 99. Some emphasis on human distinctiveness may well be necessary in order to articulate human responsibility for the destruction that has already been wrought. It seems important in this context to employ a variety of images for the human role instead of asking a single image to bear all the weight for human responsibility and care.

73. Chrétien, *Call and Response*, 30. In this context also he stresses the "choral character" of this response and its vulnerability: "in order for it to be responsive, a voice must have *all* voices in it"; 32; emphasis his.

74. "In every manifestation of God, direct or indirect, there is adventure and advent: God comes to me, God happens to me, and I am affected by his arrival. To speak of this beauty is to speak to it also, to respond to it

and to become responsible for it, just as it means becoming responsible for my own transfiguration at its hand and thus being responsible for myself more than is really possible"; Chrétien, *Ark of Speech*, 90.

75. Obviously, the implications for how to speak of differences between creatures in more productive ways need to be explored much more fully. The link between "distinctiveness" and supposed superiority also requires fuller analysis. It might also be important to explore in exactly what ways an admission of vulnerability can be helpful for ecological action. Unfortunately, these are larger questions I cannot address in any detail in the brief space allotted here.

Creature Discomforts: Levinas's Interpretation of Creation *Ex Nihilo*
Jeffrey Hanson

1. Emmanuel Levinas, *Totality and Infinity*, trans. Alphonso Lingis (Pittsburgh: Duquesne University Press, 1969).

2. As Levinas himself writes on page 28 of the Preface to *Totality and Infinity*, "We were impressed by the opposition to the idea of totality in Franz Rosenzweig's *Stern der Erlösung*, a work too often present in this book to be cited. But the presentation and development of the notions employed owe everything to the phenomenological method." The potential influence of Rosenzweig on the question of creation will be occasionally addressed in later notes, but for now the most crucial point to observe is that for both Levinas and Rosenzweig creation is a primarily moral concept; Rosenzweig argues that creation is only "*ex nihilo*" when humanity is fully moralized. See Rosenzweig, *The Star of Redemption*, trans. William W. Hallo (South Bend: University of Notre Dame Press, 1985), 119–20: "The creation of the world need reach its conclusion only in its redemption. Only from that vantage point, or wherever such an end is to be posited, would the world be—and from there, it is true, it would absolutely have to be—retrospectively a '*creatio ex nihilo*.'" Drew M. Dalton also remarks upon the critical neglect of this theme in Levinas scholarship: "Levinas' subtle resort to an account of creation is one of the most ignored, certainly one of the least explored, and perhaps one of the least understood elements of his immanent thought on the ethical life and solicitation of the Other. Only a very small handful of scholarly articles have been written on the subject and when addressed in other forms of

secondary literature, its treatment has more often than not been scant and thin at best"; see Dalton, "The Pains of Contraction: Understanding Creation in Levinas through Schelling," *Studia Phaenomenologica* 6 (2006): 215–40, 215.

3. Levinas, *Totality and Infinity*, 89.

4. See Levinas, *Totality and Infinity*, 293: "Theology imprudently treats the idea of the relation between God and the creature in terms of ontology."

5. See Dalton, "Pains of Contraction," 220–21.

6. See Levinas, *Totality and Infinity*, 293: "One may speak of creation to characterize entities situated in the transcendence that does not close over into a totality."

7. See Levinas, *Totality and Infinity*, 119: "The breach of the totality that is accomplished by the enjoyment of solitude—of the solitude of enjoyment—is radical. When the critical presence of the Other will call in question this egoism it will not destroy its solitude. Solitude will be recognized in the concern for *knowing*, which is formulated as a problem of origin—inconceivable in a totality. To this problem the notion of causality can bring no solution, since it is precisely a question of a *self*, a being absolutely isolated, whose isolation causality would compromise by reinstating it in a series. The notion of creation alone will be commensurate with such a question, respecting at the same time the absolute novelty of the I and its attachment to a principle, its having been called in question."

8. See Levinas, *Otherwise than Being*, trans. Alphonso Lingis (Pittsburgh: Duquesne University Press, 1981), 182: "Essence, the being of entities, weaves between the incomparables, between me and the others, a unity, a community (if only the unity of analogy), and drags us off and assembles us on the same side, chaining us to one another like galley slaves, emptying proximity of its meaning."

9. See Levinas, *Totality and Infinity*, 292: "The concept of a Good beyond Being and beyond the beatitude of the One announces a rigorous concept of creation, which would be neither a negation nor a limitation nor an emanation of the One." Jill Robbins adroitly addresses the complexity of Levinas's handling of the idea of participation in her *Altered Reading: Levinas and Literature* (Chicago: University of Chicago Press, 1999), 87–90. Her conclusion, that in general "the term also serves polemically

in Levinas's discourse to denote a philosophical climate from which he wishes to depart," (89) is sound. I think it is likely that in this matter as in many others, as Levinas himself cheerfully conceded in the preface to *Totality and Infinity* (see note 2 of this chapter), he is following the influence of Rosenzweig's *Star of Redemption*. Rosenzweig also explicitly opposes his notion of creation to "generation" (135–36) and "emanation" (136–37).

10. Levinas, *Totality and Infinity*, 104–5; see also 293: "To affirm origin from nothing by creation is to contest the prior community of all thing [sic] within eternity, from that philosophical thought, guided by ontology, makes things arise as from a common matrix. The absolute gap of separation that transcendence implies could not be better expressed than by the term creation, in that the kinship of beings among themselves is affirmed, but at the same time their radical heterogeneity also, their reciprocal exteriority coming from nothingness. One may speak of creation to characterize entities situated in the transcendence that does not close over into a totality."

11. Levinas, *Totality and Infinity*, 103.

12. Ibid.

13. Ibid.

14. Ibid.

15. For the meaning of this locution, see ibid., 11, 12, 120, and 261. The exact phrase "infinition of the infinite" occurs in Levinas, *Otherwise than Being*, 93, where it is equated with "glory."

16. Levinas, *Totality and Infinity*, 104; translation modified. The temptation to see in these passages a covert reference to the Lurianic kabbalist notion of the *tzimtzum* is irresistible. As Jacob Meskin notes, Levinas himself in a 1982 interview claimed that these words evoke the *tzimtzum*, but Meskin gives a sober and careful appraisal of just how to understand this reflection; see Meskin, "The Role of Lurianic Kabbalah in the Early Philosophy of Emmanuel Levinas," *Levinas Studies: An Annual Review* 2 (2007): 49–77. Michael Fagenblat finds Meskin's study somewhat persuasive but asserts that "at the very least the emphasis of these studies [that link Levinas and Luria] seems to me in need of correction. Levinas openly distanced himself from the Kabbalah, seems only to have read one Kabbalistic work (the eighteenth-century digest of Hayim Volozhin, *Nefesh ha-Hayyim*), and derided the Hasidic popularization of the Kabbalah";

see Fagenblat, *A Covenant of Creatures: Levinas's Philosophy of Judaism* (Stanford, Calif.: Stanford University Press, 2010), 202n4. Fagenblat also argues that "the comparison [between Levinas and Luria] is on the whole misplaced. . . . Levinas and Luria cannot be reconciled because Luria is a monist for whom the cause of all being and the end of all being are one, whereas Levinas is a monotheistic moral dualist for whom evil is separate and other than God" (61) and that "Levinas was opposed to the unitive mystical speculation of the Kabbalah, no doubt because of his hatred of theodicy. In his view, the desire to unify good and evil within one ultimate concept, as the Kabbalah invariably does, is both metaphysically and morally mistaken" (60). Dalton cites the above passage as his primary textual support for arguing that Levinas's view of creation entails the *contractio dei* espoused by Schelling; Dalton, "Pains of Contraction," 218. In my view commentators would do well to heed Fagenblat's caution and consider the likelihood that whatever Kabbalistic or Schellengian elements might be discernible in Levinas's thinking are almost certainly filtered through the profound influence of Rosenzweig.

17. Levinas, *Totality and Infinity*, 103; see also 105: "What is essential to created existence is not the limited character of its being, and the concrete structure of the creature is not deducible from this finitude. What is essential to created existence is its separation with regard to the Infinite. This separation is not simply a negation. Accomplished as psychism, it precisely opens upon the idea of Infinity."

18. Levinas, *Totality and Infinity*, 103–4; see also 195: "The terms, the interlocutors, absolve themselves from the relation, or remain absolute within relationship."

19. Ibid., 104.

20. Ibid., 63.

21. Ibid., 144.

22. Ibid., 103–4: "In saying that infinity is *thought* concretely starting with the separated being toward it we are nowise taking a thought that starts with the separated being to be relative. Separation is the very constitution of thought and interiority, that is, a relationship within independence."

23. Fagenblat, *Covenant of Creatures*, 102.

24. Levinas, *Otherwise than Being*, 125.

25. Ibid., 105.

26. Fagenblat, *Covenant of Creatures*, 102–3.

27. Levinas, *Otherwise than Being*, 113.

28. Fagenblat, *Covenant of Creatures*, 93.

29. Levinas, *Totality and Infinity*, 306; Fagenblat, *Covenant of Creatures*, 93.

30. More effectively than any thought, the child in her embodied and unpredictable person surely puts the lie to the Marquise de Pompadour's blithe dismissiveness, referred to in Levinas, *Totality and Infinity*, 145. No parent can say "after us the deluge."

31. Levinas, *Totality and Infinity*, 28n2: "In broaching, at the end of this work, the study of relations which we situate beyond the face, we come upon events that cannot be described as noeses aiming at noemata, nor as active interventions realizing projects, nor, of course, as physical forces being discharged into masses. They are conjunctures in being for which perhaps the term 'drama' would be most suitable, in the sense that Nietzsche would have liked to use it when, at the end of *The Case of Wagner*, he regrets that it has always been wrongly translated by action"; see Friedrich Nietzsche, *The Case of Wagner*, in *Basic Writings of Nietzsche*, ed. and trans. Walter Kaufmann (New York: Modern Library, 2000), 630n: "It has been a real misfortune for aesthetics that the word *drama* has always been translated 'action.' It is not Wagner alone who errs at this point, the error is world-wide and extends even to the philologists who ought to know better. Ancient drama aimed at scenes of great *pathos*—it precluded action (moving it *before* the beginning or *behind* the scene). The word *drama* is of Doric origin, and according to Doric usage it means 'event,' 'story'—both words in the hieratic sense. The most ancient drama represented the legend of the place, the 'holy story' on which the foundation of the cult rested (not a doing but a happening: *dran* in Doric actually does not mean 'do')."

32. Levinas, *Totality and Infinity*, 279.

33. Ibid., 109.

34. Ibid., 28n2.

35. Surely it is no accident that the dwelling in Section II is inhabited by a domestic couple, but there are no children present.

36. Levinas, *Totality and Infinity*, 277.

37. Ibid., 279. The translation of "pas seulement" in both of these passages might be better rendered as "not merely" or "not simply."

38. Ibid., 162: "Property alone institutes permanence in the pure quality of enjoyment, but this permanence disappears forthwith in the phenomenality reflected in money. As property, merchandise, bought and sold, a thing is revealed in the market as susceptible of belonging, being exchanged, and accordingly as convertible into money, susceptible of dispersing in the anonymity of money."

39. Ibid., 156–68.

40. Ibid., 272.

41. As Fagenblat rightly notes, this command is at the heart of the Judaic moral legacy and is the companion to the prohibition on murder. Murder is an affront to the literal image of God in the human being; childbearing is the literal perpetuation and dissemination of the image of God; see Fagenblat, *Covenant of Creatures*, 88–94.

42. Levinas, *Totality and Infinity*, 264–65: "The relationship established between lovers in voluptuosity, fundamentally refractory to universalization, is the very contrary of the social relation."

43. Ibid., 271.

44. Ibid., 266.

45. Ibid., 293–94.

46. Ibid, 183: "To be in oneself is to express oneself, that is, already to serve the Other. The ground of expression is goodness. To be καθ αυτο is to be good."

47. A pithy phrase from the end of the book may underscore this point too: "But then the Other . . . resembles God"; ibid., 293. This suggests that the analogue to the creator is the Other, not the subject.

48. Levinas, *Existence and Existents*, trans. Alphonso Lingis (Pittsburgh: Duquesne University Press, 2001), 57.

49. Ibid., 5.

50. Ibid., 57.

51. Ibid., 2.

52. Levinas, *Otherwise than Being*, 114.

53. Ibid., 113.

54. Ibid., 113–14.

55. Ibid., 176.

56. Ibid., 113.

57. Ibid.

58. Ibid., 118.

59. Ibid., 116.

60. Ibid., 114.

61. Levinas, *Existence and Existents*, 2.

62. Levinas, *Otherwise than Being*, 92. The phrase "en deçà" occurs at least once in *Totality and Infinity* in connection with creation; see 85 and accompanying note: "Knowledge as a critique, as a tracing back to what precedes freedom, can arise only in a being that has an origin prior to its origin—that is created." The French reads, "Le savoir comme critique, comme remontée en deçà de la liberté—ne peut surgir que dans un être qui a une origine en deçà de son origine—qui est créé."

63. Ibid., 115. Lest there be any doubt that it is the same dynamic from *Existence and Existents* now being revisited in *Otherwise than Being*, we should observe the explicit appeal to the language of hypostasis in the prior chapter on recurrence (language that does not figure into the argumentation of *Totality and Infinity*), as well as footnote 7 to chapter, which reads, "We continue to use the term *essence*, underscored, as an abstract noun of action for being as distinguished from entities in the amphibology of being and entities"; ibid.,194n7.

64. Jacques Derrida, "Violence and Metaphysics," in *Writing and Difference*, trans. Alan Bass (Chicago: University of Chicago Press, 1978).

65. Levinas, *Otherwise than Being*, 159.

66. Ibid., 157.

67. Ibid., 158.

68. Ibid., 58, 159; see also 161: "The way leads from responsibility to problems. A problem is posited by proximity itself, which, as the immediate itself, is without problems."

69. John Milbank makes this point in an unpublished article entitled, "The Shares of Being or Gift, Relation and Participation: An Essay on the Metaphysics of Emmanuel Levinas and Alain Badiou," available at http://theologyphilosophycentre.co.uk/online-papers/.

70. The family after all is the one human institution associated directly with creation by scripture—the one thing about creation that is "not good," God says in Genesis, is that man is alone, so "male and female he created them."

71. Levinas, *Otherwise than Being*, 92.

72. Ibid., 14: "As a substitution of one for another, as me, a man, I am not a transubstantiation, a changing from one substance into another, I do not shut myself up in another identity, I do not rest in a new avatar."

Reflections from Thoreau's Concord
Edward F. Mooney

1. Ralph Waldo Emerson, *Early Lectures: 1838–1842* (Cambridge, Mass.: Harvard University Press, 1972), 29: "that which is with him constitutes place. He is place and whatsoever is not with him in spirit is abroad and vagabond." On creation as our place, see Bruce V. Foltz, *The Noetics of Nature: Environmental Philosophy and the Holy Beauty of the Visible* (Bronx, N.Y.: Fordham University Press, 2014).

2. This essay was originally read at a meeting of the Society for Continental Philosophy and Theology at Loyola Marymount in Los Angeles: hence my informal references to colleagues in the audience and to the markedly un-Thoreauvian setting in Southern California.

3. "Walking," in *The Essays of Henry D. Thoreau*, ed. Louis Hyde (New York: North Point Press, 2002), 172.

4. "I am not a man, I am dynamite"; Friedrich Nietzsche, *Ecce Homo: "Why I Am a Destiny,"* trans. Duncan Large (Oxford: Oxford University Press, 2007), 88.

5. Nietzsche, "Schopenhauer as Educator," in *Untimely Meditations*, ed. Daniel Breazeale, trans. R. G. Hollingdale (Cambridge: Cambridge University Press, 1997), 129.

6. Hannah Arendt, *Love and Saint Augustine* (Chicago: University of Chicago Press, 1996), 52.

7. Henry David Thoreau, *Journal*, February 23, 1860, many editions; for instance, *I to Myself:An Annotated Selection from the Journal of H. D. Thoreau*, ed. Jeffrey S. Cramer (New Haven: Yale University Press, 2007), 425.

8. Paul J. Griffiths, *Intellectual Appetite* (Washington, D.C.: The Catholic University of America Press, 2009), 131.

9. Thoreau, *Cape Cod* (New York: Thomas Crowell, 1961), 123. See my discussions in Edward F. Mooney, *Lost Intimacy in American Thought: Recovering Personal Philosophy from Thoreau to Cavell* (New York: Continuum, 2009), 6, 13–15, 195–96.

10. The concluding line of Melville's *Moby Dick* (many editions).

11. Victor Hugo wrote a passionate plea for Brown's life from political exile; see Mooney, *Lost Intimacy*, 219.

12. Theodicy is vindication of God against the evidence of massive evil. The need to vindicate might be understood as the need for morale in the face of evil, the need to bolster one's will to live in the face of affliction.

Existentially, Job's question, "Why does God rain evil upon me?" springs from the same desperation as his wife's question to her husband, "Why not curse God and die?" Nietzsche held in *The Birth of Tragedy* that only art can redeem life in the face of such wild and pervasive afflictions. Thoreau might say that only gratitude for life can redeem it. This second framing of "the problem of suffering" allows for a solution—a redemption of life— without requiring that the afflicted assume the extravagant posture of accusing God, putting him on trial. It can accomplish the aim of classical theodicy without affirming (or denying) the standard theistic assumptions of a good, all-powerful, and omniscient deity. It is a theodicy that opens its doors to theological agnostics and unbelievers.

13. J. M. Coetzee has his protagonist describe herself as a "wounded animal" in Coetzee, *Elizabeth Costello* (New York: Viking, 2003); see the extended discussion in Stanley Cavell, Cora Diamond, John McDowell, and Ian Hacking, *Philosophy and Animal Life* (New York: Columbia University Press, 2009).

Creation and the Glory of Creatures
Janet Martin Soskice

1. James L. Mays, "'Maker of Heaven and Earth': Creation in the Psalms," in *God Who Creates*, ed. William P. Brown and S. Dean McBride (Grand Rapids: Eerdmanns, 2000), 75.

2. Ibid.

3. See Jose Faur, *Golden Doves with Silver Dots* (Bloomington: Indiana University Press), 18ff.

4. Gerhard May suggests it.

5. See Menahem Kister, "*Tohu wa-Bohu*: Primordial Elements and *Creatio ex Nihilo*," *Jewish Studies Quarterly* 14 (2007): 229–56.

6. "In regard then to this matter, which is so great and wonderful as to be sufficient for all the bodies in the world, which God willed to exist . . . I cannot understand how so many distinguished men have supposed it to be uncreated, that is, not made by God himself the Creator of all things, but in its nature and power the result of chance"; *Origen on First Principles*, trans. G. W. Butterworth (London: S.P.C.K., 1936), 79. Origen may just be addressing pagan philosophical opponents, of course.

7. This is the view of Gerhard May, who writes that Hellenistic-Jewish philosophy expressed the omnipotence of God and his role as creator, but

"did not engage in a fundamental debate with the Platonic and Stoic doctrine of principles. And so it could not develop a doctrine of *creatio ex nihilo*. The statements about creation of non-being or from non-being . . . are not to be understood as antitheses to an eternal matter and to the principle of '*ex nihil nihil fit*' but to be considered as an unreflective, everyday way of saying that through the act of creation something arose which did not previously exist. As soon as it was freely recognised that creation by the biblical God was more than the forming of matter, that he brought forth the world in sovereign freedom and without any external conditions, the expression 'Creation out of nothing' offered itself as a formula pregnant to describe the particular character of the biblical concept of creation"; May, *Creatio ex Nihilo*, trans. A. S. Worrall (Edinburgh: T. and T. Clark, 1994), 21.

8. I would say, too, from later middle- and neo-Platonists; see below.

9. On this see Paul Gavrilyuk, "Creation in Early Christian Polemical Literature: Irenaeus against the Gnostics and Athanasius against the Arians," *Modern Theology* 29, no. 2 (2013): 22–32, and the excellent article by Paul M. Blowers on "Doctrine of Creation," in *The Oxford Handbook of Early Christian Studies*, ed. Susan Ashbrook Harvey and David G. Hunter (Oxford: Oxford University Press, 2008).

10. May speaks of the "decisive basis of the doctrine of *creatio ex nihilo*, that the biblical conception of God demands that matter also be created"; May, *Creatio ex Nihilo*, 15.

11. Some pagan philosophers of this period also believed Plato had learned from Moses. John Dillon writes, "Philo may well have been acquainted with that part of the Pythagoras legend that we find evidence of much later in Iamblichus," *Life of Pythagoras* (14), according to which Pythagoras spends a while in Palestine "consorting with the descendants of *Mochos* the prophet and philosopher"; Dillon, *The Middle Platonists* (London: Duckworth, 1977), 143.

12. In this Philo and the Jews of Alexandria were by no means alone. In Philo's time Greek was the working language of the eastern Roman Empire. Paul's native language was Greek, and there were Greek-speaking cities in Judaea and Galilee. By the time of Philo, Paul, and the New Testament writings, "Greek rhetoric, philosophy and literature were the stable of elite education," including that of elite Jews; see Tessa Rajak, *Translation and Survival: The Greek Bible of the Ancient Jewish Diaspora* (Oxford: Oxford University Press, 2009), 94–95.

13. May, *Creatio ex Nihilo*, 10.

14. Ibid., 15, 12.

15. Dillon, *Middle Platonists*, 155.

16. See David Runia's commentary to Philo, *On the Creation of the Cosmos according to Moses*, trans. Runia (Atlanta, Ga.: Society of Biblical Literature, 2005), 132–33.

17. See David Runia, "Plato's *Timaeus*, First Principle(s), and Creation," in *Plato's Timaeus as Cultural Icon*, ed. Gretchen J. Reydams-Schils (Notre Dame, Ind.: Notre Dame University Press, 2003), 137.

18. "How must it not be impossible to recompense or to praise as He deserves Him who brought the universe out of non-existence?" (*L.A.*, III.10); "He is full of Himself and sufficient for Himself. It was so before the creation of the world, and is equally so after the creation of all that is. He cannot change nor alter and needs nothing else at all, so that all things are His but He Himself in the proper sense belongs to none" (*De Mut.*, IV.27); "Through His goodness He begat all that is, through His sovereignty He rules what he has begotten" (*De Cher.*, 27–28); see also *Fug.*, 46, *Moses*, II.267.

19. Philo, *On the Creation of the Cosmos according to Moses*, introduction, trans., and commentary by David T. Runia (Atlanta: Society of Bibical Literature, 2001), 157.

20. Dillon, *Middle Platonists*, 155.

21. Hebrew scholars take Exodus 3:15 to be saying that the Tetragrammaton is the name "for all generations." Elsewhere Philo, like Aquinas, identifies "He that Is" (*ho on*) as serving as God's proper name (Cher., 27 ff). There is no reason, of course, why more than one "licensed" name could not serve in this way. None of them, from Philo's point of view, is adequate.

22. St. Augustine, *De Trinitate*, Book V, Prologue.

23. See Carol Harrison, *Rethinking Augustine's Early Theology* (Oxford: Oxford University Press, 2006), chap. 4.

24. On this, see Rowan Williams, "God for Nothing: Augustine on Creation," *Augustinian Studies* 25 (1994): 9–24.

25. Augustine, *Confessions*, trans. Maria Boulding (London: Hodder and Stoughton, 1997). Chadwick translates "in peace . . . the selfsame. . . . I will go to sleep and have my dreams."

26. I have written about Augustine on these matters in more detail in Janet Martin Soskice, "Augustine on Knowing God and Knowing the Self," in *Faithful Reading: New Essays in Theology and Philosophy in Honour*

of Fergus Kerr O.P., ed. Simon Oliver, Karen Kilby, and Tom O'Loughlin (London: T. and T. Clark, 2012), 61–74.

27. This is suggested by Miles Burnyeat in his "Platonism in the Bible: Numenius of Apamea on *Exodus* and Eternity," in *The Revelation of the Name YHWH to Moses: Perspectives from Judaism, the Pagan Graeco-Roman World, and Early Christianity*, ed. George H. van Kooten (Leidin: Brill, 2006); see especially 158–60.

28. Similar glosses are found in other Targumim.

29. Robert Louis Wilken, *The Spirit of Early Christian Thought* (New Haven: Yale University Press, 2003), xvi.

Care of the Soil, Care of the Self: Creation and Creativity in the American Suburbs
T. Wilson Dickinson

1. Ellen Davis, *Scripture, Culture, and Agriculture: An Agrarian Reading of the Bible* (New York: Cambridge University Press, 2009).

2. Ibid., 127–29.

3. Ibid., 10.

4. Charles Taylor, *Modern Social Imaginaries* (Durham: Duke University Press, 2004), 23–25.

5. Michel Foucault, *The Hermeneutics of the Subject* (New York: Palgrave-Macmillan, 2005), 10, 11.

6. Jean-Louis Chrétien, "The Wounded Word," in *Phenomenology and the "Theological Turn": The French Debate* (New York: Fordham University Press, 2000), 154.

7. Martin Heidegger, *Being and Time*, trans. Joan Stambaugh (Albany: SUNY Press, 1996), 110–18.

8. Michael Fishbane, *Sacred Attunement: A Jewish Theology* (Chicago: University of Chicago Press, 2008), 18.

9. Heidegger, *Being and Time*, 111.

10. Norman Wirzba, *The Paradise of God: Renewing Religion In an Ecological Age* (New York: Oxford University Press, 2003), 2.

11. Michael Pollan, *Second Nature: A Gardener's Education* (New York: Delta, 1991), 75; Paul Robbins, *Lawn People: How Grasses, Weeds, and Chemicals Make Us Who We Are* (Philadelphia: Temple University Press, 2007), 33–44.

12. Wirzba, *Paradise*, 76.

13. Robbins, *Lawn People*, xii.

14. Ibid., 114.

15. Paulo Freire, *Pedagogy of the Oppressed* (New York: Continuum, 1989), 58.

16. Ibid., 60.

17. Wirzba, *Paradise*, 86.

18. Catherine Keller, *Face of the Deep: A Theology of Becoming* (New York: Routledge, 2003), 6.

19. Ibid., 26.

20. Simone Weil, *Waiting for God* (New York: Perennial Classics, 2001), 62.

21. Janet Martin Soskice, "Love and Attention," in *The Kindness of God: Metaphor, Gender, and Religious Language* (New York: Oxford University Press, 2007), 32.

22. Fishbane, *Sacred Attunement*, 42.

23. Wirzba, *Food and Faith: A Theology of Eating* (New York: Cambridge University Press, 2011), 50.

24. Ibid., 57.

25. Pollan, *Second Nature*, 76.

26. Robert Pogue Harrison, *Gardens: An Essay on the Human Condition* (Chicago: University of Chicago Press, 2008), 74.

27. Wirzba, *Paradise*, 31.

28. Wirzba, *Food*, 52.

29. Weil, *Waiting*, 62.

30. Wirzba, *Food*, 54.

31. Søren Kierkegaard, *Upbuilding Discourses in Various Spirits* (Princeton: Princeton University Press, 1993), 162.

32. Karmen MacKendrick, "The Hospitality of Listening: A Note on Sacramental Strangeness," in *Phenomenologies of the Stranger: Between Hostility and Hospitality*, ed. Richard Kearney and Kascha Semonovich (New York: Fordham University Press, 2011), 105.

33. Ibid., 103.

34. Ibid., 105, 106.

35. Wirzba, *Paradise*, 191.

36. Ibid., 114.

37. Keller *Face of the Deep*, 230.

Dream Writing Beyond a Wounded World:
Topographies of the Eco-Divine
Susan Pyke

1. The interrelationship between habitat and habits has long been debated in ecological sociology and psychology. Peter Khan's work on nature and technological nature points toward the positive relations between humans and their physical environment; see Khan, Rachel L. Severson, and Jolina H. Ruckert, "The Human Relation with Nature and Technological Nature," *Current Directions in Psychological Science* 18, no. 1 (2009): 37–42. John Thøgersen has been working for some years on the potential correlation between pro-environmental attitudes and behaviors, as least insofar as this relates to "spillover" actions and policy support. His work with Tom Crompton can be found in Thøgersen and Crompton, "Simple and Painless? The Limitations of Spillover in Environmental Campaigning," *Journal of Consumer Policy* 32, no. 2 (2009): 141–63. Nonetheless, the gap between attitudes and behaviors remains, and is well documented in social science. The work of Anja Kollmuss and Julian Agyeman remains seminal; see their "Mind the Gap: Why Do People Act Environmentally and What Are the Barriers to Pro-Environmental Behavior?" *Environmental Education Research* 8, no. 3 (2002): 239–59. I thank Jim Curtis for these insights.

2. Emily Brontë, *Wuthering Heights*, ed. Pauline Nester (1847; repr. Middlesex: Penguin, 2003), and Anne Carson, "The Glass Essay," in *Glass, Irony, and God* (New York: New Directions, 1995).

3. Grace Jantzen formalizes this term in her exploration of Luce Irigaray's work, including Irigaray's poetic essay from which I take my meanings for divine; see Irigaray, "Divine Women," in *Women, Knowledge, and Reality: Explorations in Feminist Philosophy*, ed. Ann Garry and Marilyn Pearsall (New York: Routledge, 1996). Jantzen's title is just the beginning of a wonderful analysis of Irigaray's work; see Jantzen, *Becoming Divine: Towards a Feminist Philosophy of Religion* (Bloomington and Indianapolis: Indiana University Press, 1999).

4. As Lynn White Jr. has pointed out, this perspective can be found in the teachings of St. Francis of Assisi. It is also present in the thinking that makes up the "green" perspective of Ecumenical Patriarch Bartholomew; see White, "The Historical Roots of Our Ecologic Crisis," *Science* 155 (1969): 42–47, and John Chryssavgis, "Ecumenical Patriarch Bartholomew:

Insights into an Orthodox Christian World View," *International Journal of Environmental Studies* 64, no. 1 (2007): 9–18.

5. John Grim describes his term through what he has learned of Aboriginal and Torres Strait Islanders' diverse and "deep, attractor relationships between people and place that activate the affective, cognitive and creative forces at the heart of cultural life" in his essay "Indigenous Knowing and Responsible Life in the World," in *Ecospirit: Religions and Philosophies for the Earth*, ed. Laurel Kearns and Catherine Keller (New York: Fordham University Press 2008), 200. As with my project, Grim is not attempting to assimilate these ways of thinking in his exploration of "lifeways," but rather observes and learns from the specific relations these communities have with country.

6. Freya Mathews, *Reinhabiting Reality: Towards a Recovery of Culture* (New York: SUNY Press, 2005), 76.

7. This definition comes from various readings, but most of all I am indebted to Deborah Bird Rose, who offers the base of this definition in her work, which includes *Nourishing Terrains: Australian Aboriginal Views of Landscape and Wilderness* (Canberra: Australian Heritage Commission, 1996).

8. I use Cory Wolfe's definition here, where the post-human involves both the "thematics of a decentralizing of the human" through evolution, ecology, and technology and a thinking that "confronts that thematic" where "the nature of thought itself" changes to the post-human; Wolfe, *What Is Posthumanism?* (Minneapolis: University of Minnesota, 2010), xvi.

9. Again, I owe this conceptualization to Rose (*Nourishing Terrains*), who owes her summary to the communities that have shared their knowledge with her.

10. See John Ruskin's *Modern Painters*, ed. David Barrie (1907; repr. London: Deutsch). However, as Rigby points out, Ruskin does allow for "the impingement of the environment on human subjectivity"; see Kate Rigby, *Topographies of the Sacred: The Poetics of Place in European Romanticism* (Charlottesville and London: University of Virginia Press, 2004), 80.

11. From Onno Oerlemans, *Romanticism and the Materiality of Nature* (Toronto: University of Toronto Press, 2001), 21–23.

12. Rigby uses the phrase "two-way dynamic of place and perception" (*Topographies of the Sacred*, 13) and goes on to eloquently describe the role played by Romanticism in thinking country in this way. There have been some important evolutions. As eco-critic James McKusick points out,

at its inception, the word "ecology" suggested a household run by human-kind, a place contained and ordered and limited by human boundaries of understanding; McKusick, *Green Writing: Romanticism and Ecology* (New York: St. Martin's Press, 2000), 20. Emerging terms, such as biophilia (see Khan et al, "Human Relation," for more on Edward Wilson's use of this term) suggests something more inclusive and coaffective.

13. Rigby, *Topographies of the Sacred*, 77.

14. Ibid., 158.

15. Ibid., 162.

16. Enid L. Duthie, *The Brontes and Nature* (Houndmills, Basingstoke and London: Macmillan), 238.

17. Ibid., 329.

18. Ibid., 49.

19. Ibid., 130.

20. Ibid., 3–4.

21. Irigaray, *Marine Lover of Friedrich Nietzsche*, trans. Gillian Gill (New York: Columbia University Press, 1991), 16.

22. The "male," Irigaray would argue, is the "human" here, for tra-ditionally "woman becomes a place for man"—"she becomes a thing." Humans' not-male potential lies somewhere in the night of post-human possibilities; Irigaray, *An Ethics of Sexual Difference* (1984; repr. London: Continuum, 2004), 11.

23. Ibid., 43.

24. David Abram, *The Spell of the Sensuous: Perception and Language in a More-Than-Human World* (New York: Vintage, 1997), 274.

25. Irigaray, *Marine Lover*, 130.

26. This is, as Judith Butler might call it, an "emerging and unpredict-able assemblage," where a shared orientation to the eco-divine is marked by an acceptance of alternatives beyond the normative Ideal Man; Butler, *Gender Trouble: Feminism and the Subversion of Identity* (1990; repr. New York: Routledge Classics, 2006), 20.

27. Ibid., 82.

28. As Bracha Ettinger describes it, this coaffectivity echoes the inter-uterine experience of being more than one. The enclosed self is released into a partial subjectivity that coexists with the (m)other; Ettinger, *The Matrixial Borderspace* (Minneapolis and London: University of Minnesota Press, 2006).

29. Ibid., 113.

30. For Carson's speaker, "turning into Emily Brontë" involves walking in all weather and communing with a "Thou" read more from the Book of Nature than the Book of Scripture. In this "turning," Carson's speaker suggests Brontë heard from country through her mind and body, then voiced its affect through her poetry and novel; Carson, "The Glass Essay."

31. As Cixous puts it, "in writing [or reading] our body is alive: we either use the whole of it or, depending on the dream, a part. We must embark on a body-to-body journey in order to discover the body"; Hélène Cixous, *Three Steps on the Ladder of Writing* (New York: Columbia University Press, 1993), 65.

32. Ibid., x.

33. The potential productivities in this line of thinking can be found in Freya Mathews's idea of narrative invitation. "If Dreaming is the subjectival dimension of the world, then it seems that we can engage this Dreaming dimension by creating a narrative context in which it can express itself to us. This is our invocational task, the task of singing up. Our invocations will, of course, have to be phrased in the poetic language of things as opposed to the conceptual language of words, since the world does not think, at least in any literal way, in the conceptual language of words. This poetic language is however familiar to us from dreams, where meaning is conveyed predominately through objects and circumstances rather than through discourse"; Mathews, "The World Hidden within the World: A Conversation on Ontopoetics," *Trumpeter Journal of Ecosophy* 23, no. 1 (2007): 76.

34. Carl Jung, *Memories, Dreams, Reflections*, ed. Aniela Jaffé, trans. Richard Winston and Clara Winston (London: Collins, 1963), 163, 257, and 218.

35. Decentering the human runs counter to the prevailing culture, and living against the grain can be frightening. Little wonder not all readers welcome the kind of dream writing offered by topographies that listen for and respond to place. The idea of sentient place only makes sense after being given a good talking-to by backchatting country determined to be heard.

36. Mathews, *Reinhabiting Reality*, 128.

37. Ibid., 21–22.

38. Gilles Deleuze and Félix Guattari, *A Thousand Plateaus: Capitalism and Schizophrenia*, trans. Brian Massumi (London and New York: Continuum, 1987), 7, 9, 25.

Contributors

T. Wilson Dickinson is Associate Dean for Religious Life and Adjunct Assistant Professor of Religion and Philosophy at Transylvania University. He is currently completing a book entitled, *The Wisdom of the Cross: Scandalous and Transformative Exercises from Corinth to Paris.*

Bruce Foltz is Professor of Philosophy at Eckerd College in St. Petersburg, Florida, and Founding President of the International Association for Environmental Philosophy. He is the author of *Inhabiting the Earth: Heidegger, Environmental Ethics, and the Metaphysics of Nature* (Humanities Press) and *The Noetics of Nature: Environmental Philosophy and the Holy Beauty of the Visible* (Fordham University Press), as well as coeditor of *Toward an Ecology of Transfiguration: Orthodox Christian Perspectives on Environment, Nature, and Creation* (Fordham University Press). He is finishing a new monograph to be called *Nature and Other Modern Idolatries.*

Christina M. Gschwandtner teaches Continental Philosophy of Religion at Fordham University. Her publications include *Reading Jean-Luc Marion: Exceeding Metaphysics* (Indiana University Press, 2007); *Postmodern Apologetics? Arguments for God in Contemporary Philosophy* (Fordham University Press, 2012); and *Degrees of Givenness: On Saturation in Jean-Luc Marion* (Indiana University Press, 2014).

Jeffrey Hanson is Research Fellow in Philosophy at Australian Catholic University. He is the editor of *Kierkegaard as Phenomenologist: An Experi-*

ment and coeditor with Michael R. Kelly of *Michel Henry: The Affects of Thought* (Bloomsbury, 2013).

Jarrod Longbons is a Ph.D. candidate at the University of Nottingham, as well as one of the pastors of the historic Peachtree Christian Church in Midtown Atlanta, Georgia. His research focus is primarily on the doctrine of creation and its relationship to ecology.

Edward F. Mooney is retired from the Departments of Religion and Philosophy, Syracuse University, and is Visiting Professor at Tel Aviv University and Hebrew University of Jerusalem. His most recent books are *Excursions with Kierkegaard: Others, Goods, Death, and Final Faith* (Bloomsbury, 2012) and *Lost Intimacy in American Thought: Personal Philosophy from Thoreau to Cavell* (Bloomsbury, 2009). Earlier books include *Selves in Discord and Resolve: Kierkegaard's Moral Religious Psychology* (Routledge, 1996) and *Knights of Faith and Resignation: Reading* Fear and Trembling (SUNY Press, 1991).

Susan Pyke teaches at the University of Melbourne, where she recently completed her Ph.D. in Creative Writing. Her critical publications focus on the interaction between literature and ecology, and she has also published poetry, fiction, and associative essays. Her most recent work can be found in *Southerly*, the *Australasian Journal of Ecocriticism and Cultural Ecology*, and *Text* (Special issue, "Writing Creates Ecology and Ecology Creates Writing").

Janet Martin Soskice is Professor of Philosophical Theology at the University of Cambridge and, since October 2014, President of Jesus College, Cambridge. Her recent books include *The Kindness of God* (Oxford University Press, 2008) and *Sisters of Sinai* (Chatto and Knopf, 2009). She is an editor, with David Burrell, Carlo Cogliati, and Bill Stoeger, of *Creation and the God of Abraham* (Cambridge University Press, 2010).

Brian Treanor is Professor of Philosophy and Director of Environmental Studies at Loyola Marymount University. His work applies hermeneutics to a wide variety of topics and issues, including embodiment, place, literature, ethics, religion, and environmental themes. He is the author, most recently, of *Emplotting Virtue* (SUNY Press, 2014) and the coeditor of *Interpreting Nature* (Fordham University Press, 2014).

The Very Reverend Rowan Williams, formerly the 104th Archbishop of Canterbury, is now Master of Magdalene College at the University of Cambridge. He is the author of multiple books, including *On Christian Theology* (Wiley-Blackwell, 2000); *The Wound of Knowledge* (Cowley, 2003); *Wrestling with Angels: Conversations in Modern Theology* (SCM Press, 2007); and *Dostoevsky: Language, Faith, and Fiction* (Baylor University Press, 2008). He delivered the Gifford Lectures in the fall of 2013.

Norman Wirzba is Professor of Theology and Ecology at Duke University's Divinity School and Research Professor of Theology and Ecology at Duke's Nicholas School for the Environment. He is the author of *The Paradise of God: Renewing Religion in an Ecological Age* (Oxford University Press, 2007); *Food and Faith: A Theology of Eating* (Cambridge University Press, 2011); and, most recently (with Fred Bahnson), *Making Peace with the Land: God's Call to Reconcile with Creation* (IVP Books, 2012).

Index

gROUNDWORKS|

ECOLOGICAL ISSUES IN PHILOSOPHY AND THEOLOGY

Forrest Clingerman and Brian Treanor, series editors

Interpreting Nature: The Emerging Field of Environmental Hermeneutics
Forrest Clingerman, Brian Treanor, Martin Drenthen,
and David Utsler, eds

*The Noetics of Nature: Environmental Philosophy
and the Holy Beauty of the Visible*
Bruce V. Foltz

*Environmental Aesthetics: Crossing Divides
and Breaking Ground*
Martin Drenthen and Jozef Keulartz, eds.

The Logos of the Living World: Merleau-Ponty, Animals, and Language
Louise Westling

Being-in-Creation: Human Responsibility in an Endangered World
Brian Treanor, Bruce Ellis Benson, and Norman Wirzba, eds.